SKY GODS & ANGELS

OR

OUR ANCESTORS RETURN FROM SPACE TO EARTH

Wence Horak

CRUX EDITION

OTHER BOOKS BY THE AUTHOR:

ANCIENT ECOLOGISTS   (an e-book and a pb)

NOSTRADAMUS DECODED   (an e-book and a pb)

Published by
Earth Way Society
Cover image: courtesy NASA
306-1349 Bertram Street
Kelowna, B.C. V1Y 8N4
Canada

ISBN 978-0-9688889-6-4 (e-book)
ISBN 978-0-9916741-4-5 (paperback)

CONTENTS

FOREWORD

"The linear concept of history is unsustainable."
Prof. Josef Polisevsky

A child can accept that God resides in heaven, the enormous expanse of sky overhead, made magical during the day by the cloud animals, floating and endlessly shifting there, by the warmth of the glowing Sun, the rain or snow that descend from it, by the brilliance of the stars that shine there at night, by the mystery of the waxing and waning Moon.

However, when the child grows up into a young adult who finds out that heaven is virtually synonymous to a firmament, the young adult is puzzled, even troubled, wondering why would a God and his angels, all presumably insubstantial, need a firmament up there. The puzzlement increases when the young adult finds out the "angels" used to father children with mortal women. That seemed odd, to say the least, and it presented the young mind with another puzzle, another mystery to solve.

Attempting to deal with these puzzles, to unravel the mysteries, I traveled many a beaten path, as well as the untrodden ground, opened doors I was shoved in front of, as well as those hidden in ruins, behind brambles. Strangely enough, the highlights of the knowledge I picked up here and there eventually started to dovetail each other. Or rather fit each other like pieces of an ancient puzzle.

Eventually, I found out there was more than one firmament in the sky, and that there was also more than one God. Even the Bible hints at it: "And the woman said unto Saul, I saw Gods ascending out of the Earth." (Samuel XXVIII:13) Also, "O Lord our God, other lords besides thee have had dominion over us …" (Isaiah XXVI:13) Other mythological sources speak of the gods a lot more openly.

In fact, myths from all over the world speak of flesh and blood beings living in the sky, and visiting the planet, as if it were a commonplace event – sometimes a very, very long time ago. And whether the relevant stories originate in the Middle East, or North America, they tell us the sky-dwellers were having children with the men and women of Earth, just like the angels. That also means these people were no fly-by aliens, whose separate evolution would have made them totally unsuitable for such results, but had to be natives of this planet, and indeed, many myths confirm this directly or indirectly.

If all the myths of this world were to be gathered in books, they would fill hundreds of volumes. Having studied them practically all his life, Joseph Campbell justifiably considered myths to have a central role in human history. Claude Levi-Strauss also correctly observed that "Throughout, myths resemble one another, to an extraordinary degree." Another anthropologist, Clyde Kluckhorn, agrees that "There is an astounding similarity between myths collected in different regions," and he adds: "What is puzzling ... (are) the identical themes from different cultures ... repeating images and symbols."

Until now this puzzle remained unanswered, because the mythographers and anthropologists have failed to satisfactorily answer the question of what was behind all these identical themes. And they have failed to do so, because they have limited themselves only to the sociological or psychological interpretation of the myths.

Then, in 1969, Giorgio de Santillana and Hertha von Dechand published their magisterial *Hamlet's Mill*, in which they demonstrated that many of the ancient myths have their roots in astronomy and mathematics.

While certainly daring in their willingness to rock the boat of preconceptions, the two authors have still failed to notice two other unifying themes that permeate all major mythological systems. Namely the ecological and climatic themes, which are discussed in my book *Ancient Ecologists,* and the technological themes, which are the subject of this book.

As you will find out though, the two themes became very closely linked in the past. Those of you who are perhaps familiar with the subject of paleoclimatology will know that climate is very changeable, over the long periods of time. Also, all the major climactic changes bring with them a total devastation for the planet, for the people who live on it, and the civilizations they may have developed.

For example, a Persian 'myth' claims that only 2,900 people survived the cataclysmic ending of the last Ice Age. The Canadian environmentalis David Suzuki estimates only 10,000 people have survived a similar event, some 100,000 years ago. In fact, all major mythological systems claim a total of five cycles of civilizations have preceded this one. All of them, as will be demonstrated, have been destroyed by climatic upheavals, starting about one million years ago, or by events that precipitated them.

When the leading members of one of these pre-historical civilizations realized how fragile and uncertain the existence of a civilization is on this planet, they had decided to take pretty drastic action. Having reached a technological level superior to ours, they concluded it would be much easier to construct huge habitats in the space above the planet and to live inside these habitats. Some of the 'myths' quoted here actually describe the various stages of this feat, and other relevant details of their construction.

It will be a 'wild ride' so, please hold onto your hats and senses. The reward will be a much better understanding of our (distant) past. But having a much better grasp of the reality of our past will enable us to better project our future, and thus make it much safer, saner, and more enjoyable, by avoiding the fateful mistakes, and the vain glory of the past.

TELLTALE SIGNS

"I am the God of the sky's spaces. I originated time and form ...
I created the divine material which reproduces itself ... I am Aton."
Egyptian Book of the Dead

Proving that an advanced civilization, indigenous to our planet, and technologically vastly superior to ours, had evolved here, is not as impossible to prove as it might at first seem – in spite the lack of material evidence. The telltale signs were always there, but instead of being systematically collected, such evidence was being systematically ignored. There were several reasons for that, one of them as 'simple' as being brainwashed into believing the current version of human evolution.

Let us first consider the knowledge of atoms. The earliest recorded awareness of these building blocks of matter is about 2,500 years old, and pops up virtually simultaneously in Greece and India. Democritus wrote: "In reality, there is nothing but atoms and space." About the same time an Indian sage Uluka said that all material objects were made of 'seeds' of matter. In the *Yoga Vasishta* we could read: "There are vast worlds within the hollows of each atom." This sentence also hints at the existence of subatomic particles. The *Vahamira Table*, also about 2,500 years old, even mathematically indicates the size of an atom.

A few centuries later the Roman author Lucretius stated in his book *On the Nature of Things*: "But things are made of atoms; they are stable. Until some force comes, hits them hard, and splits them ... " And here, of course, we are already talking nuclear fission. It was not until the 20th century that scientists became aware of such a possibility, which did not become reality until World War II.

But the awareness of the forces governing the physical universe ran deeper than that. Much deeper. In one of the Hindu sacred books, *The Laws of Manu*, it is stated: "But joining minute particles even to those six, which possess measureless power, with particles of himself, he created all beings."

While this might sound like another confused, bombastic statement by one of the ancient seekers, and there were many of those, we also have to realize there indeed are three key pairs of quantum particles that started the formative process of the material universe. There is also the Higgs boson, known as the God particle, that gave the particles mass. Yet, the awareness of the formative forces goes to the state even before the initial stages of the universe, also recorded in the *Laws of Manu*: "(The) universe existed in the shape of Darkness, unperceived ... then the Divine Self-Existent ... appeared with irresistible creative power, dispelling the darkness ... First with thought (He) created the waters, and placed his seed in them. That seed became a golden egg, in brilliancy equal to the Sun..." ("In that egg he himself was born as Brahman, the progenitor of the whole world.")

Here, foremost, we have to try to clarify the concept of 'water', so often used in ancient stories of creation. The best explanation I came across is to be found in the story of Cosmic Creation, of the Zuni Indians of New Mexico: "In the beginning of things Awonawilona was alone. There was nothing beside him in the whole space of time. Then Awonawilona conceived in himself the thought, and the thought took shape ... (and) stepped out into the void, into outer space, and from them came the nebulae of growth and mist, full of power of growth. [T]he nebulae condensed ... and became water. "

In addition to the fact both of the above 'myths' agree with the basic tenet of our own sciences, specifically astrophysics, about the origin of our universe, and of the empty void that preceded it, the Zuni version makes it quite plain that it was also understood, in a very distant past, that the nebulae consisted mainly of hydrogen, the most common gas in the universe, which in its 'concentrated' form, as H_2O, becomes water. Unfortunately, the idea of gases concentrating into the future solar systems got somehow confused with the idea of gases concentrating into water.

Secondly, we have to clarify here the distinction between the ancient ideas of the earliest stages of the birth of the universe, which are more akin to the concepts of quantum mechanics and strings theory, rather than to the still prevalent ideas about an "accidental creation" of the universe. Needless to say, I am inclined to side with the ancient view, because, as we shall see later, in much greater detail, their knowledge was simply superior to ours. It also included the concept of a Big Bang, albeit conceived very differently, as can be noted in the segment from the *Ethiopic Enoch*, in which the Lord, we are told, explains the initial stages of creation.

" ... I have decided to lay foundation – to cause a visible universe. I ordered in the underworld for one of the invisible ones to descend as a visible one. Adoel descended, exceedingly large. I looked at him, and behold, in his stomach he had a great light ... and he opened up and mighty light shone forth, and I was in the midst of the light. As the light floated, behold, a great age issued from it, revealing the entire creation ..."

This narrative seems to suggest that the primeval creative, self-aware entity, consisting of sub-quantum particles, or energy so ethereal we cannot detect it, (re: the mysterious 'black energy'), transformed a part of self into the visible universe, (re: Brahma), which, in turn, is a part of the multiverse. ("… one of the invisible ones …") This concept can also help to explain the puzzle of the matter 'missing' from our universe.

The next stage of creation, as preserved in myths and sacred books, tells us about the formation of our solar system – preceded though by the 'death' of Chaos, according to a Chinese source. A Taoist story reveals its demise was brought about by the kindness of two rulers: Shu (Heedless) and Hu (Hasty). Yet Hu can also be translated as "Sudden Energy," i.e. a Big Bang. Obviously, the Taoist jesters in their ignorance transformed the Supreme Being into the character of Shu. On the other hand, another Chinese source claims: "There is order at the core of the greatest disorder," which mirrors the key idea of the modern theory of chaos.

The Mayan *Popol Vuh* describes the formation of the solar system this way: "Like the mist, like a cloud, and like a cloud of dust was the creation," preserving the idea that solar systems are formed from the accumulation of space dust and gases. Anaximander, in the 6th century Before Current Era, or BCE, stated, presuming the Earth to be larger than the Sun: "Sun, Moon and the stars were created by a *centrifugal* force from the *flaming* surface of the Earth." While certainly flawed in details, at least this statement preserves for us the essential idea that the solar system was formed out of the flaming, disc-shaped body of the premordial Sun, by the centrifugal force.

Anaximander also discusses the concept of "apeiron," an unlimited, immortal "naturalness," out of which "excrete" the opposites, and eventually an enormous flaming ball forms, which splits into numerous fiery rings, out of which in turn the Sun, Moon and stars were created. The story of the flaming hot origin of our solar system was preserved by the Shastika Indians of California as well. They 'remember' that "In the beginning, Sun had nine brothers, all flaming hot like himself." Interestingly enough, while the current solar system contains only eight planets, not counting Pluto, a ninth planet is said to have existed, situated between Mars and Jupiter, where numerous planetoids are located nowadays, in the so-called asteroid belt. Furthermore, the next stage of the process, that is of a cooler 'crust' forming atop the still hot planet, is correctly 'remembered' by the Innuit, who say the Earth "Already had a firm crust before there was any land."

As far as the position and behaviour of the Earth within the solar system, and in turn of the solar system within the universe, these too were also understood once, as the following statement and quotes can attest to. The aforementioned Anaximander wrote: "The Earth is round and it revolves around the Sun." About three centuries later Aristarchos, according to Plutarch, stated: "The Earth revolves in an oblique circle around the Sun, while it rotates at the same time about its own axis." Heraclides of Pontus was even more specific on the topic, saying "The Earth spins on its axis once in twenty-four hours." And the ancient Hindu book *Surya Siddhanta* is said to contain reasonably accurate calculations of the diameter of Earth and its distance from the Moon. The Sumerians knew the Earth's radius to be 6,310 kilometers, or a difference of mere 1%. Today we know it to be 6,371 kilometers. Yet, since our planet is 'pear' shaped, both figures could apply. Erasthenes also provided us with a fairly accurate circumference of the Earth in third century BCE. The fact the Moon shines with light borrowed from the Sun was also known to the ancient Greeks and Hindus, as was the 'mystery' of solar occlusion. Anaxagoras stated, some 2,500 years ago, "It is the Moon that darkens the Sun during an eclipse." The same gentleman also said: "The Sun is a vast mass of incandescent metal." While, strictly speaking, this statement is not correct, it too contains a kernel of truth, because any Sun ultimately produces vast amounts of heavy elements, including metals, when it explodes.

To conclude this section of the chapter, we should point out that the ancient people were even aware of the age of our solar system, which they rendered as the Day of Brahma, said to be 4.32 billion years, while contemporary scientists estimate the age of our solar system to be about 4.5 billion years. By now we also know this solar system is but one of many. The ancient people knew that as well. Democritus wrote some 2,500 years ago: "Space is filled with myriads of stars and the Milky Way is but a vast conglomerate of distant stars." Some 200 years later Aristarchos of Samos added: "The distances that separate us from stars are immeasurable." In Cicero's model of the universe the Earth realistically occupied a very minor position. He also claimed immense stars existed in the Milky Way, and the Egyptians used to say there were millions of millions of stars in the universe, even though only eleven thousand are visible in the sky. Of those, 1,464 were catalogued by the ancient Chinese and only slightly more by Tycho Brahe in the 17th century. And finally, Lucretious wrote: "Cosmos is full of worlds like our own, where life manifests itself in various ways." In other words, the ancients most likely knew a thing or two we still do not.

The incidents of parallel knowledge between the ancient and modern sciences even include examples of the very evolution of life on this planet. For starters, Zuni Indians describe this event: "The Sun-Father formed the seed-stuff ... impregnating therewith the great waters, and lo!, in the heat of his light these waters of the sea grew green, and scums rose upon them, waxing wide and weighty ... " The Phoenicians creation myth also spoke of "The putrification of watery mixture, and from this sprang all the seed of creation."

The 'slime', of course, is the premordial blue-green algae that eventually produced the first amounts of oxygen. And Neb-er-cher, the Egyptian primal supreme God, made this claim: "I evolved myself out of the primeval matter, which had evolved multitudes of evolutions from the beginning of time." With a bit of liberty we can say Neb-er-cher, in general, is Life itself. The rest is quite self-explanatory. For those who need a hint, we only have to point out that several evolutionary waves swept over this planet during the last roughly 600 million years, following the emergence of the 'slime', whose 'seeds' had arrived here from space.

Anaximander offered that the first living creatures were born in "damp places," and that their bodies changed after coming on dry land, and shedding their *prickly skin*.

Such a statement, until fairly recently, would have been deemed rather nebulous. Yet the specific reference to the "prickly skin" helps us to shed some light on this statement. Presumably most readers have heard of the famous Cambrian explosion of life, some 600 million years ago, or more specifically of life that was able to leave behind hard shells. Here we have in mind the shells found at the site of the Burgess Shale, which are really notorious for their fantastic, almost artistic shapes, most of which are 'prickly'.

It would seem that the Mande people of Africa have remembered, for the rest of us, the fact that birds are really descended from giant flying reptiles. This idea is linked with a folk hero named Kassa Kena Gananina, who was extremely powerful. One day a monstrous dragon-bird of the kind called Konoba came to hover above him, and he killed it with his iron staff. The monster fell dead, but one of its feathers landed across the hero's shoulders, struck him down, and half buried him in the ground. The feather was so heavy, even his friends could not budge it. After a while, an old woman, most likely representing the evolutionary process, came by, carrying a child on her back. Stooping down, she softly blew upon the feather until it gently drifted away. Then she picked up the slain Konoba, by then an ordinary bird, giving it to the little child to carry.

Lucretius notes on related subject: "In those days, again, many species have died out altogether and failed to reproduce their kind." (*On the Nature of the Universe.*) Alexander Elliot, in his book *The Universal Myths*, brings our attention to the "Common conception in myth ... of two mountains ... one of which points up and the other down. The bases of the two mountains meet and are joined at the Earth's surface." Yet, only some ten years ago a group of geologists announced the existence of 'shadow mountains' under the surface of our planet, reflecting, one could say, the mountains above. And, of course, the subterranean mountains point downward from a joint base at the surface!

In this context, it should not be at all surprising that people would have retained even a vague memory of the DNA, the basic 'building material' of life. We come across the idea in the German origin myth, presenting us with a god Twisto, i.e. "two-fold," who is born of the Earth, and has a son, Mannus, who becomes the originator of the human race. While it is a standard practice to explain away Twisto as an androgenous, or two-sexed deity, we could just as well turn this logic around and deem all the androgenous gods to represent the DNA. After all, as we shall see in the next chapter, the myths have retained for us also the memory of genetic labs as well.

If we were to sum up the details of the ancient knowledge, as preserved by our ancestors, we would realize their genesis is almost identical to our genesis: Initially, there was but an endless, 'empty' void, and then the Universal Mind called for the beginning of the material universe. First, there was a sudden surge of quantum particles, which resulted in a tremendous 'fireball'. In the next stage, these particles organized into atoms, of which hydrogen, the lightest and the most plentiful is predominately mentioned, albeit as 'water'. Inevitable confusion, as well as a 'rule' of most myths, reverses the order of events, because the nebulae did not condense into 'water', but rather the hydrogen (a.k.a. 'water') condensed into nebulae. The formation of the first generation of stars is, again, understandably skipped, or forgotten, except perhaps as the idea that stars are "made of iron." But the formation of our solar system out of an immense cloud of cosmic gas and dust was preserved in the memory, as was the idea of the flaming hot disc, which gave birth to the Sun and the planets. Even the formation of cooler crust, on the surface of the still hot Earth was remembered, as was the origin of life out of 'slime', which we politely call blue-green algae, and the scientists named them "prokaryotes." The 'myths' also remember something about the first living creatures on this planet, about the DNA, even about the long process of evolution, of the many waves of it, on this planet. And they used to be equally aware of the age of our solar system, of its relevant position within the universe. It should come as no surprise then that some 2,500 years ago Heracleitos of Ephesus proclaimed "This world ... was not created by any God ..."

Obviously, such knowledge could not be obtained without very advanced sciences. By the same token, such knowledge could not have been imparted to primitive humans by a group of visiting aliens, because the primitives simply would not be interested in it enough to commit it to a long-range memory, represented by the myths. Besides, the aliens would have to teach the same thing all over the globe, not to speak of the fact they most likely would not know about the Burgess Shale or that they would not have bothered digging for dinosaurs, specifically the pterodactyl's bones, and study in detail the evolution of life here. All of that during a brief exploratory visit that would also have to coincide with the cataclysm that ended the Ice Age proper, some 15,500 years ago and assist with the ecological salvage of the planet, and, and ...

As we shall see, there are just too many 'ands' to make the alien cop-out tenable. Like it or not, we simply will have to accept the idea that, once, there used to be a very advanced civilization, technologically superior to ours, and all that is left of it, as far as we know, are a few bits of information scattered amidst and buried under mounds of distorted, misunderstood, misinterpreted, and 'edited' gobbledygook, either in oral or written form.

That such a civilization could not have come into existence during the prehistoric period following the termination of the last Ice Age ought to be evident from a remark in one of the Persian myths which claims only 2,900 people survived the cataclysm of 15,500 years ago, when the planet was devastated by mega tidal waves and mega hurricanes.

Whatever the number of survivors might have been, they would have been numerically too small to create any civilization at all. In fact, as the myths inform us, the survivors either became cannibals or worse, they sunk to a subhuman level, crawling naked on all fours. Having lost the gift of speech, they were unable to communicate, and thus to form even the most basic form of culture, or society.

Yet, as if on a cue, the sky gods descend, and within a few millennia set everything right. It was also these 'gods' who provided the survivors with language, and other rudiments of culture, as there was essentially only one language and one culture on the planet, back then, as the experts tell us. Gradually, they imparted more and more knowledge to the again ever-growing population, but the subsequent cataclysms that included three 'lesser' global floods, and three asteroid impacts, played havoc with the re-acquired knowledge.

Then, during the so called "historical times," there were also massive population shifts, and wars for living space that resulted in more destruction and more forgetting. In the Roman Empire, Carthage and China, millions of irreplaceable books were burned, entire libraries destroyed, many of them, without a doubt, containing very relevant information. We only know of one such book, *Sphere*, about Earth, written by a priest of the Eleusinian Mysteries, named Musaeus, some 3,400 years ago.

It is not generally recognized, but in the ancient time the priesthood was more often associated with the safeguarding of knowledge than anything else. Unfortunately, this all started to change since about 4,000 years ago, after the 'last' global flood, and the 'last' asteroid impact. This decline is also associated with the disappearance, or fall of the gods. (Re: the 'fall of angels' in the Bible, or the departure of the kachinas, in Hopi myths.)

However, this event was very real, as demonstrated in chapter The Fall.

From that time on all the material evidence, specifically of the technological type, gradually kept disappearing from the surface of the planet, but not necessarily the human ability to produce technologically interesting items.

TECHNOLOGY AND SCIENCE IN MYTHS AND HISTORY

Anyone remotely interested in the mysteries of the past must be familiar with the images of the prehistoric astronauts. Perhaps the earliest ones that became known were from Tasilli, in the Sahara. The next most widely recognized became the Dogu figurines from Japan. Yet, the likeness of such spacemen can also be found drawn on the cave walls in the Swiss Alps, in Central Asia, and Australia.

A similar reference to a space-suited being was preserved for us in the Mesopotamian myth of Oanaes, the 'fish-god', who came out of the sea. He is portrayed as having a second head under the fish one, human feet, but a tail of a fish. Since Oanaes, after emerging from the sea, had spent an entire day on land instructing the people, the fish tail most likely was merely an in-suit-toilet, similar to the one used by modern-day astronauts. In the story of the Gaelic Argonauts, we can read that in order to reach the Island of Fianchuive, which no boat or ship could reach, "Brian left his brothers and put on his water dress, with his transparency of glass, upon his head." But, as we shall see here shortly, this Brian most likely arrived in such a suit from the said 'island'.

The few sentences the Encyclopedia Britannica devoted to the Hyperboreans mention the fact that most of them, after they have arrived in their ship at the Island of Delos, have died. According to the Gaelic lore, most of the gods who came to live in Ireland, namely the Partholon and the Nemed people, have died after encountering outsiders. If such a fate reminds you of the fate of the North American Indians, who died after an exposure to new viruses introduced by the Europeans, it certainly is no coincidence. However, since the stories are quite plain about the one-sided impact of such encounters, we have to assume that neither the Hyperboreans, nor the Partholons, or the Nemeds, infected the other side. It also means they all had arrived from a preciously clean environment, free of any viruses, a place which the Gaelic story calls the Other World.

As we shall see ever more clearly, they indeed did come from a world completely separate from this one, while at the same time also a part of it, but free of any viruses or diseases – and for as long as they could, they had tried protecting themselves from such, while on the planet, by using either germ suits or germ screens. And several myths can attest to their existence.

Perhaps the simplest, and the most non-committal, is the Maidu Indian story of the Earth-Initiate, who came down from the sky via a rope of feathers. His face was covered and never seen, but his body shone like a Sun. Of course, most germ suits, even today, are made out of plastic, and shine. The person's face remained invisible because the visor was either opaque, or too dark to see through, like a one-way mirror, or the visor used by jet pilots. Furthermore, the Earth-Initiate was most likely one of the sky people, specific to the Maidu. Such sky people, according to the Amerindian and other myths, used to come down male and female, and 'married' the surface people, as they used to refer to them. Many of the marriages came about after the sky people, usually the sky maidens, had their 'bird-costumes' stolen, while having a bath in a lake or a river.

The Innuit story on the same subject is much more interesting. It is about a long lost daughter who came to visit in a sled pulled by invisible dogs. After the arrival "she told her mother that her husband said to put a seal gut curtain in front of their sleeping place. She said they would not stay if they did not put up a curtain." Further on the story explains that, the man, (i.e. the husband), was from the Moon. Interestingly enough, the parents could see their daughter, but could only see the man's shadow against the curtain. The visit over "they loaded their sled … (it) went on the ground for a while, then they went up in the air."

In the Sumerian myth of Enki and Ninhursag we are told, in a very wobbly fashion, that a god, identified as Enki, provided mankind with various fruits and vegetables – by applying genetic engineering. As the text states, he "gave fate to all plants, came to know them completely." The knowledge of genetics is further confirmed by an Indonesian myth about a sky dweller named Ratu Champa. He descended to Java, where he practiced, in secret, his high craftsmanship. He concealed his creations in a mountain cavern, the cavern door he kept carefully shut, for everything inside it had an *independent life*. Of course, viruses, germs, DNA, and manipulated molecules of living matter, tend to have an independent life, which is why the modern genetic and similar labs are also 'carefully shut' by reverse pressure, lower on the inside, to prevent the research material from escaping.

From the Chinese sources we learn about the Fiery Emperor, also known as the Divine Peasant, because he provided the people with the five cereals, and taught them how to plant them. He is also known for giving plants their healing properties. Some stories claim he was poisoned seventy times in a single day while testing them, some say he died of it, and one of them even claims he had a glass window on his stomach to observe the process, and was able to neutralize the poisons. In general, we are being presented again with the results of genetic/medical research, and specifically with the use of a microscope.

The same country provides us with a story of a man who was in possession of eight silver statues that were able to walk. One day these statues walked out of the house, turning into young men who could talk. If such a brief report can make you think of (human-shaped) robots, you are on the right track, as Homer mentions two types of them in the *Iliad*. First of all, there are the Hephaestus' handmaids, "all cast in gold, but a match for living, breathing girls. Intelligence fills their hearts, voice and strength their frames ..." In addition to these 'decorative' items, Homer also mentions the much less glamorous work-type robots, explaining that Hephaestus was "pressing the work of twenty three-legged cauldrons, an array to ring the walls inside his mansion. He'd bolted golden wheels to the legs of each so all on their own speed, at a nod from him, they could roll to halls where the gods convened, then roll right home again ..." It is also said of the smelting cauldrons of the famous engineer Yü, who lived in the 24th century BCE in China, that they "moved of themselves, could boil without being heated ..." (I.e. without a visible fire.)

However, while such references are not entirely unique, since later on we shall introduce a small army of specialized robots from a Samoan myth, we ought to point out over here that virtually every culture producing metal tools and weapons had its own Hephaestus, or divine blacksmith who, more often than not, was also associated with medicine and other forms of sacred knowledge. In Egypt he was known as Ptah, and all his priests in fact used to be engineers, or scientists of one kind or another, as well as doctors and surgeons, until the steep decline, following two major cataclysmic events in the late third millennium BCE.

We can also add here that Hephaestus is credited with creating Talos, the huge monster, (i.e. robot), who is said to have guarded Crete, after it was abandoned, circa 1,500 BCE. Most likely the strange character of Humbaba, guarding the gods' forest in the *Epic of Gilgamesh*, was also a robot. One is even said to have been in the service of Egyptian priests in the Temple of Ra, until the time of Alexander the Great.

Robots, of course, are technologically inseparable from computers. The ancients used to call them Teraphim, or animated 'stones', first mentioned by the Phoenician historian Sanchuniathon, around 1,200 BCE. The Christian historian Eusebius (260 – 340 CE) carried one of them on his chest. It answered questions put to it in a small voice reminiscent of a low whistling, we are told. Another source speaks of a "clear and sharp small voice." We also know that Moses allegedly gave his brother Aaron two such stones. These prophetic, or translation stones, were usually reserved for high priests. As the philosopher Maimonides (1135-1204 CE) comments: "It is in this manner that the Teraphim taught people many useful acts and sciences."

The myths and ancient fables even describe an automated flying medical robot that we know under the name of the Golden Fleece. Unfortunately, its existence and its function can only be extrapolated from the properties associated with it and its function, and from the name of Medea herself, so inseparably linked with the Golden Fleece.

If you will recall, it was Jason and his fabled band of heroes, including Hercules, sailing aboard the ship Argo, who retrieved, (i.e. stole), the Golden Fleece from Colchis, situated by the historians somewhere on the south-eastern shores of the Black Sea, with the help of Medea, the local princess.

Those who read the story or saw the film version of it hopefully realize it is nothing more than an improbable fable. It is to me, anyhow, especially in its historicized garb. But if we peek under the simplistic surface of it, an entirely different view of ancient events starts emerging. When we consider the name of the ship itself, the ARG part refers, in the ancient Indo-European language, to a white metal, possibly steel. The GO part most likely referred either to a skin, or a horned animal, such as ox, bull, or cow. Whether we decide it originally meant a 'steel-clad' or 'the bull of steel,' will hardly make any difference.

Then we ought to point out the aerial combat between the Harpies, called "the Hounds of Zeus," and the sons of Boreas, stationed aboard the Argo. It would in fact seem that Argo was an aircraft carrier, of sorts. An obviously related tale speaks of the Hyperboreans chasing the Harpies twice around the globe in an aerial combat. Yet no ordinary plane could fly that far. However, we could pursue various parts of the story, with rather unsatisfactory results, only to realize it was clobbered together from many discordant parts, most of them hoary with age, even when these stories were written down.

At the heart of the story though, is Medea. That name, according to my Indo-European dictionary, is comprised of two words. The "med" part means to take appropriate measure, or to look after, to heal, and it wound up in English as the word medication. The "dea" part refers to a divinity. Combined, the word Medea means a "divine healer." Needless to add, her character has undergone the usual negative reversal suffered by most ancient deities. Yet, some of her original aspects are still retained in the story, namely her ability to "make old folks young," or rather to maintain their youthful vigor.

After her 'evil deeds' were done, Medea, on the roof of Jason's palace, stepped into a chariot drawn by dragons, to be carried away through the air. But we have to realize that Medea, the flying chariot, and the Golden Fleece, are one and the same. After all, with Medea gone we no longer hear of the Golden Fleece, which was nothing else but a flying medical robot. It is also quite possible that the machines, once they started to break down, probably killed, or otherwise harmed the would-be patients. (Re: Medea's 'evil deeds'.)

On the other hand, a so-called Hyperborean priest from Britain arrived to visit Pythagoras in Sicily, flying atop a device much like the Golden Fleece, and this visit is also associated with medical treatments. We also need to add that even the name Jason meant a "healer."

After this detective detour, we can again return to the more direct references of advanced technology in pre-historic times, concentrating now perhaps on the reports of flying machines. For example, in the Babylonian set of laws called the Halkatha, it is stated: "To operate a flying machine is a great privilege. Knowledge of flying is most ancient, a gift of the gods of old for saving lives." (Re: the Golden Fleece.) Also, the Chaldean text known as *Sifr'ala*, dating before 3,000 BCE, though fragmentary, fills almost 100 pages in English. It provided a detailed account of how to build and operate an aircraft, naming parts such as graphite rods, copper coils, vibrating spheres, and comments on wind resistance, gliding and stability, but is too fragmentary to actually reconstruct a plane.

A large store of literature still exists in India, describing various flying machines, their construction and parts, as well as the means of their propulsion. There is also the documented story of Bartholomeu de Gusmao, a Portuguese Jesuit priest, who had been taught in Bolivia, South America, how to build a plane. He actually built one with a grant from King John V of Portugal, and successfully flew it in 1709. Then the Inquisition stepped in. The plane was destroyed, the plans and additional volumes of notes taken to the Vatican. These notes also included knowledge of constructing devices for lifting great weights and a space-going vehicle.

A great many reports of flying machines reach us from China as well. Historically, the oldest is probably the one about Emperor Shun, who is said to have reigned between 2258 and 2208 BCE. Reportedly, he not only constructed a flying craft, but also tested a parachute. Then, in 1776 BCE, Emperor Cheng Teng ordered a court artisan Ki-Kung-Shi to construct a flying apparatus. He did and flew it to the province of Honan. The Emperor had it destroyed to prevent it from falling into the wrong hands.

Such stories though pale in comparison to the memory of spaceships and space travel. According to the Chinese historical tradition, Hou Yih, the engineer of Emperor Yao, traveled to the Moon aboard a celestial bird that mounted a current of luminous air, in 2,309 BCE. This sounds very much like using a propulsion rocket. Another story from the same time period speaks of an enormous ship that appeared on the sea at night, brilliantly lit, able to travel to the Moon and the stars.

Both of the most famous Hindu epics, *Mahabharata* and *Ramayana*, mention aerial vehicles known as vimanas, that apparently varied in size and shape, and were able to soar above the clouds, and to fly into space. Some of them were described as having two decks, some with sides of iron and clad with wings. The latter also seems to fit the picture of the "silver birds" that transported the Innuit to their new home in the Arctic, from the heartland of the Asian continent. All in all, hundreds of stories exist in India on the topic of a "sky-coach," or of a "metal horse" in the sky. Models of planes, or gliders, were found in South America and Egypt. Furthermore, the ancient Greek playwright Aeschylus mentions people who constructed wheel-less vehicles with linen wings to carry them over trackless waters, in his play *Prometheus Unbound.*

The myths even remember a form of movie-house projection effect, as well as a television. In the Australian aborigines story about Winjarning brothers, who were captured by the Keen Keengs, a race of winged, but otherwise humanoid demiurges. They made the brothers sit "with their backs to a fire which leapt up from a pit. They witnessed ... ritual dancing abound the fire pit, but could see only the shadows which were cast by the dancers on the wall in front of them. They heard the singing clearly and memorized it."

The situation described above not only made me think of educational movies, but also strongly reminded me of Plato's famous "cave of ideas." As a philosopher, he imagined mankind sitting in a cave with its back to the entrance, catching only glimpses of the real world outside, the world of ideas, as shadows on the wall of the cave. He was partially correct, of course, but I still suspect he got his visual idea from a source similar to the one we know from Australia.

The Chinese story The Trials of Langoan tells us that after he was unfaithful to his wife she showed him a looking glass which contained the image of him and the other woman making love. A Czech fairytale describes an entire picture gallery of prospective brides, in motion, and in living colour. Solomon was also credited with owning such a seeing mirror, as were the Olympian gods.

Magical sources of light not only were mentioned in myths, but some physically survived until recent times. For example the lamp that illuminated Noah's Ark was called Tsohar. In literal translation the word means "shining crystal." The Sumerian flood-hero Utu brought his rays into the boat to give it light. In the ancient Jewish manuscript *The Queen of Sheba and Her Only Son Menyelek,* translated by Sir E.A. Wallis Budge, it is stated: "Now the house of Solomon the King was illuminated as by day, for in his wisdom he had made shining pearls which were like unto the Sun, the Moon and the stars in the roof of the house." Saint Augustine (354-430 CE) saw a lamp in the temple of Isis, in Egypt, that neither wind nor water could extinguish. About 700 years ago, near Rome, the sepulcher of Pallas was opened. A lantern was found inside that was illuminating the space for over 2,000 years. In 1601, in the depth of Mato Grosso, South America, the adventurer Barco Cenenera came across this scene: "On the summit of an 8 meter pillar was a great Moon which illuminated all the lake, dispelling darkness." And in Papua/New Guinea, as late as 1963, the visitors to a village of Mount Wilhemina reported "many Moons suspended in the air and shining with great brightness." Other visitors described the same Moons as huge stone balls that began to glow after sunset, according to a United Press wire.

Other sources of light were also reported. In India, around 500 BCE, the king of doctors, Jivaka, possessed a "gem," which when placed before a patient, illuminated his body as a lamp lights up all objects in a house, and so revealed the nature of his malady. Around 200 BCE, in the Palace of Hien-Yang, in Shensi, China, was found a precious mirror that illuminates the bones of the body. It measured 4 x 5 feet and gave off a strange light on both sides. The view of the body organs could not be blocked by any obstacles – as is the case with our x-rays.

The lore left behind by the Templars also mentions a peculiar metal ball. It not only produced light, but also destroyed several towns and castles in Cyprus after it was touched. Eventually it was thrown in the sea, which produced a storm – and caused all the fish to die in the affected area, never to reappear.

The Amerindians preserve for us stories of boats that moved by magical means. "But the boat moved without (paddles). It skimmed over the water by itself." Wishray Indians. "… rowed his canoe by saying 'Go fast! Go fast!' " Madoc Indians, California. "Without any paddling the canoe moved away, as if it were alive, and it traveled fast." Tsimshian Indians, North-West Coast.

And finally, in 1962, the skull of a pre-historic bison, found in Yakuzia, north-east Siberia, had a circular hole in its forehead, most likely caused by a projectile weapon. Also, the Neanderthal skull, found at Broken Hill, South Africa, appeared to have a bullet entry hole on one side, and was shattered at the opposite end – so typical of dum-dum bullets, such as the one that killed President John Kennedy.

Again, as we saw, the global myths have collectively preserved for us indicators of a technological society superior to ours. However, as we shall see in the following paragraphs, the ability to create technology has not exactly been absent from recorded history either. What was absent, was the will to exploit it systematically the way we do, to organize the theoretical foundation for it that we call science.

Yet, the last statement, to a degree, is incorrect, because several centuries after the massive cataclysm that ended the Ice Age proper, around 15,500 years ago, a Cankam University was established somewhere is Asia, according to the Tamil sources. So were cities, now found underwater off the coast of India and Japan. However, after its foundation the university was interrupted three times by three separate sets of lesser cataclysmic events, but re-founded only twice. The same source also comments on the ever diminishing intellectual capacity of the students in each subsequent round. While we know that universities had existed, at least in association with major temple complexes in Egypt and Mesopotamia, we are also aware that after the last double cataclysm, in 24th and 20th century BCE, all the known civilizations were restarted on a much lower intellectual level, resembling the previous cycle only in the superficial fashion. Still, the universities continued to produce scribes, clerks, accountants, doctors and architects, as well as astronomers. If any awareness of the high knowledge remained by 1,000 BCE, it was in a form of secrets entrusted to the senior priests, but science was no longer being taught.

In other words, the drive was originally there to teach science and to implement the regained knowledge, but the population base was very small, and as the population kept growing, it was being subjected to the twined effects of global floods and asteroid impacts roughly every 3,000 years, which negated the early attempts to restart an advance civilization. Still, slowly at first, things started happening.

First, a series of twelve temples, rich in symbolic art, were constructed, between 12,000 and 10,000 years ago, in Gobekli Tepe, today's Turkey. About 9,000 years ago the first 'castle' stood in the Middle-East, known as Proto-Jericho. The earliest excavated town we know of, Catal Huyuk, dates to circa 8,500 years ago. The first pyramids were actually constructed in today's Bosnia, some

8,000 years ago, at which time the first known civilization started flourishing in today's central Sahara.

Since about 6,000 or 7,000 years ago, construction methods featured earthquake-proof double walls for natural air-conditioning. Some of the architectural features of the homes for hot climate, that used to exist in Egypt, were so sophisticated modern architects were simply unable to reproduce them from description, until they found substantial remains of such a house buried in sand. The same can be said about the ecological sophistication of the Andean farmers, and their ability to create micro-climate for their many varieties of potatoes in the harsh Andean conditions.

Indoor plumbing, flush toilets and bathrooms were obviously the standard features in all middle-class homes of the Indus Valley civilization, since about 5,000 years ago. Needless to add, the cities there also featured very sophisticated sewage system. The same features were found in palace and temple complexes, as well as upper class homes in Mesopotamia, Egypt and Crete. Even in Scotland, in a place called Skara Brae, was found a drainage system for lavatories. Some of the colder places also featured central heating under the floors. While on the subject of heating, we should also mention the automated hatcheries of the ancient Egyptians, employing only the heat of the Sun to incubate eggs. From more recent times of ancient Greece and Rome, we also have evidence of showers in public baths and sport complexes and of such modern concept as condominium apartments. Even the use of cement predates the Roman period by about 4,000 years.

While the technological sophistication was absent in central Europe, of some 7,500 years ago, each family home there consisted of a spacious kitchen/living area, and of six additional rooms. The three on the left were reserved for the female members of the family – daughter(s), mother, grandmother, and those on the right for the male members. Providing about 90 square meters of living space (about 900 square feet), and being dry and warm in winter, it offered a better place to live than most New York tenements, in the opinion of one archeologist. Within a few thousand years the new standard for such families was a one room wattle and mud shack.

Since the cataclysm of 15,500 years ago, the population grew relatively slowly, although the myths have noted population explosions as well, specifically the Mesopotamian myths, and their subsequent collapse in the wake of another sudden climatic swing.

But the archeologists see the same pattern in central Europe. The first population wave started expanding there about 8,000 years ago. Within about 1,000 years the region became ecologically devastated, and most of the remaining population shifted south-east, to southern Russia and the Ukraine. The next round of population expansion into central Europe, about 6,000 years ago, also coincides (roughly) with the advent of the Age of Metals. (Yet, three copper objects, some 9,000 years old, were found in the Near East and a similar find was made in the Balkans, and dated to about 7,500 years ago.)

Finding and smelting metals is a very complex process, and as Theodore Wertime, an expert on the origin of metallurgy once wrote: "One must doubt that ... (it) could have been spun in human history." The myths of all the metal working people definitely agree on that. They all say it was taught to them by a divine smith who descended from the sky. But mainly, as Mircea Eliade points out: "A mine or an untapped vein is not easily discovered; it is for the gods and divine creatures to reveal where they lie ... These beliefs were held in European countries until quite recently."

Indeed, it would be rather difficult to imagine how the Phoenicians could, for example, without the detailed geological knowledge, and without the satellite topography, just sail across the Atlantic, then up the Mississippi, to stumble upon a rich copper mine in Michigan, as they did about 3,000 years ago. There obviously was a connection between the Phoenicians knack for trading in metals and the fact that the goddess Juno kept her 'chariot' in Carthage, as the lore claims she did.

The first known copper mine existed in the Balkans about 6,500 years ago, following the veins of chalcopyrite about 20 to 25 meters below the ground. Fairly massive copper tools, such as the combination of axe and adze, appear in the general area a few centuries later. That means ovens capable of reaching temperatures of 1,100 degrees Celsius must have been also available to melt the copper and to cast it into molds. However, there is evidence of the use of cold-hammered copper in Anatolia, (today's Turkey), some 10,000 years ago. Smaller objects, such as awls, were in general use in the Balkans almost 7,000 years ago.

The production and the use of bronze, an alloy of copper and either of tin or arsenic, started about 5,000 years ago, and virtually simultaneously in the Balkans, the Aegean, and the Near East. The tin usually was imported over fairly long distances.

One reason for the sudden blossoming of the Bronze Age is the emergence of Medzamor, near the eastern shores of the Black Sea, as the Pittsburg of the ancient world, some 5,000 years ago. The excavations indicate Medzamor was eventually producing fourteen varieties of bronze and steel, the latter in the second half of its 2,000 year long history. They also reveal the workers there not only used a variety of 'modern' tools, but also breathing apparatuses, goggles and protective clothing. Their market, to make such an operation feasible, must have included the entire arc from the Middle East to Europe, and would have included several million people. Because the regional density of the population greatly differed, as do the estimates for global population, the 'market' might have been substantially greater. Considering the conditions of the times, and the logistics involved in organizing such an 'enterprise', the achievement is truly staggering, since it also included imports of raw materials from the entire 'known world', customized production and export. Incidentally, raw iron had been sold in Europe shaped like a pyramid – its crystalline form – which means someone must have seen the crystals of iron under a microscope.

As far as the use and manufacture of iron and steel is concerned, the picture, while less clear, is more interesting. Most orthodox authorities claim the first sources of iron for weapons and tools were meteorites, and as such were used from the late period of the third millennium BCE. Yet, the Smithsonian Institute and the U.S. Bureau of Standards have brought to light objects proving that 7,000 years ago people were producing steel in furnaces at temperatures of 9,000 degrees Fahrenheit or 3,900 degrees Celsius. Furthermore, there is an indication that the iron ore mine on the island of Elba was occasionally used since about 30,000 years ago.

And talking of these far off dates, we also ought to mention the 25,000 year old hematite (iron ore) mines in Swaziland, Africa, and the ocher mine, discovered in Africa, up to 50 feet (17 meters) underground, that was in use some 130,000 years ago.

Pillars of pure iron, the largest of which weigh about six tons, were found in India and Germany. They are estimated to be about 4,000 years of age, and they are said to consist of 99.5% pure iron. Today iron of such purity is produced only in small quantities, by electrolysis. Batteries for electrolysis have certainly existed at that time, but producing such columns that way would have been costlier than making them out of gold. Incidentally, modern technologists still cannot figure out how the individual layers of these columns were so *seamlessly* 'melded' together, because they cannot imagine they could have been produced otherwise. As far as their function is concerned, we can only speculate it must have been similar to the piezzoelectric effect created by the megalith, whose purpose was to stimulate the human mind.

At the turn of our era, as far as the calendar is concerned, during the reign of Emperor Tiberius (42 BCE – 37 CE), an unfortunate Roman genius invented the (production method of) aluminum. Fearing the devaluation of gold, the Emperor had him promptly executed. During the same time period someone invented an unbreakable glass in Rome. Based on the same fear, the same Emperor had the factory destroyed.

About three centuries later, someone was luckier with the invention of aluminum in China, judging from the metal girdle general Chou-Chu was buried with in 316 CE. It contained 85% aluminum, 10% copper, and 5% manganese. But that was the last time anyone made, or saw, aluminum, until 1825, when it was produced by a chemical method. Only later, for industrial purposes, electrolysis was used.

Ancient Alexandria, with its vast library, containing hundreds of thousands of books, drew like a magnet the best minds from the Hellenistic world, and became the jewel of the Greek applied sciences. It was there Ctesibius invented a steam engine and a hydraulic pump, and Heron a steam turbine, later used on Nero's boat, as well as a mechanical singing bird. In addition to these inventions, which were not duplicated till the 18th century CE, over 100 different other automatons were also created there, similar in principle to the flying wooden dove, constructed by Plato's teacher Archytas, centuries earlier.

It was most likely also in Alexandria, around 87 BCE, that the famous Antikythera computing device was designed and made. It used fixed gear ratios, as well as a differential gear assembly, to produce calculations of the solar-lunar calendar, which were then most likely delivered in form of pointer reading on a digital dial.

Now let us turn our attention to ancient optical lenses. Until very recently I was aware of only two of them. One was found on Thera and one in Niniveh. They were both made from rock crystal, and dated before 1,500 BCE. Then I read Robert Temple's *The Crystal Sun*. According to his meticulous, but all too detailed account, over 450 ancient crystal and glass lenses are to be found in the museums – without the curators being aware of that, regardless whether the museum is in Athens or London. Heinrich Schliemann alone dug out 48 of them in Troy. There are also the Carthagian lenses, the Mycenian lenses, the Rhodes lenses, and the Ephesus lenses.

The ancient Greeks and Romans used them as spectacles, and constructed telescopes with them, as no doubt did the Egyptians and Babylonians. Lexicon, compiled in Peru by a Dominican priest Domingo De San Thomas, also describes an Inca made telescope. Closer to our times still, a telescope was constructed by Robert Bacon, in 13th century Oxford.

As far as the medical sciences are concerned, it is fairly well known the ancient Egyptians used to successfully perform cataract operations, and that they used a 'primitive' form of antibiotics. A 'primitive' form of vaccination is said to be also described in the *Vedas*. The Indian people used to practice contraception, as did the ancient Hebrews. The Amazon natives practice it till this day. In ancient Peru, quinine and belladonna were used as an anesthetic. And millennia earlier the stone-age people performed successful brain and heart surgeries.

Millions of books, contained in the libraries of the ancient world, especially between 4th century BCE and 4th century CE were destroyed, thus eliminating virtually all the remnants of the ancient knowledge, including later the Druidic Library in Carnac, France. And the Druids, according to Aristotle, Pythagoras, and others, were the foremost men of scientific knowledge of the times. In addition to the libraries that we do know that were destroyed, such as those in Alexandria, Carthage and Ephesus, (the Temple of Artemis), there must have been numerous libraries destroyed also in India and Persia, during the centuries of wars with the invading Aryans. In China, the Emperor Shinh-Huang-Ti had all the books on science and history burned in 240 BCE. The same fate befell the Mayan and Aztec books, the Inca's kipus, in 16th century.

About a 1,000 year long hiatus was enforced on technological tinkering in Europe, after the fall of Rome, the subsequent shifting of nations, and later the 'disfavour' with which the Church had viewed any such ideas. Nothing of real importance had happened anywhere else during that time, except the invention of gunpowder in China. The Arabs became known for their high quality steel swords, but otherwise they only maintained the old knowledge, including that of the Greeks. The Byzantines were equally sterile, as were the people in India.

But then, from the13th century on, some of the pre-existent ideas were being slowly 'dusted off' by the Cistercian monks in France, and later with the help of the aforementioned Roger Bacon. He also mentioned the ability of the ancients to fly. In France rabbi Jechiele, praised for erudition by King Luis IX, was said to know the secret of "a dazzling lamp that lighted itself." He also used electricity to shock unwanted visitors, according to several contemporary chroniclers.

Before we move on though, we have to mention the one bright exception, dimmed by the centuries of darkness. The exception was a French monk Gerbert, who later became Pope Sylvester II. He lived during the 10th century, and was initiated by the Arab masters in Seville and Cordoba. Later, among other things, he invented a steam organ and a weight clock. He also knew that the Earth was round, and could prove it – but no one understood him by then.

From the 15th century reports reach us of ingenious mechanical 'toys.' One of them was a fly that flew inside a room without bumping into the walls, and returned to the hand it alighted from. There is a similar report of a mechanical eagle that flew out to greet the arriving guests and lead them (back) to the castle it departed from.

Then a strange qualitative jump takes place in the 16th century. Conrad Haas, the master gunner of Sibiu, (Romania), designed and launched a multi-tiered rocket with delta wings in 1555, using solid powder fuel. This fuel, according to the report by professor Doru Todericiu, published in the Romanian Historical Review, in 1967, could have been replaced by an ethyl acetate ammoniac based chemical, presumably available at the time. One of the authenticated drawings, from 1536, portrays a three-tier rocket, with a pilot cabin on the top.

The 16th century became known as the "Century of Discoveries," starting with the 'discovery' of the Americas by Columbus, in 1492. Yet, this too likely is a case of the hen and the egg. Many people suspect he was guided in his search by a copy of an ancient map, pointing to the now famous map of the Turkish admiral Piri Reis, dated to 1513, that shows parts of the east coast of the Americas, of the Caribbean, and of the Amazon estuary, as well as the Antarctic, which was not discovered until 1818. Interestingly enough, the Land of the Queen Maud, in the Antarctic, is marked on his map by a constellation of Snake, visible only from the latitude of the said Land of the Queen Maud. In total, there were 210 of such copied maps in Piri Reis' collection, and they all used spheric trigonometry and division of the circle into 360 degrees. One of them shows the Quadalquivir river delta in Spain, but the delta itself is hardly noticeable. To grow to the contemporary dimensions would require about 20,000 years, someone calculated. The Mediterranean islands are noticeably larger on these maps, and Sweden, Germany, England and Ireland are covered in glaciers, also suggesting the situation roughly 20,000 years ago, during the peak of the glaciation. Furthermore, evidence deduced from the design of the maps suggests they were originally drawn from a point about 100 kilometers above the planet.

Two additional maps corroborate the time of origin of these maps. One of these is the so called Ptolemay's Map (of the known world), redrawn in the 15th century. It shows Greenland not quite covered in ice, but an icefield in Scandinavia, which by then was ice free. The reverse was true of Greenland. The map of Hadzhi Ahet, another Turkish admiral, drawn in 1559, is even more telling. It portrays the Arctic, in detail, as well as the Pacific coast of North America, *plus* an 'unknown' land between Alaska and Siberia. This 'unknown' land is known to paleogeographers as Beringia, and it was there throughout the last Ice Age, disappearing during the cataclysm of the 15,500 years ago, when the sea level rose by some 150 meters.

However, Columbus did not necessarily need the ancient maps to surmise a continent exists on the other side of the Atlantic Ocean from Europe. For example, the Roman author Plutarch wrote the following: "in the midst of the Western Sea lies Ogygia, the Island of Venus and Calypso. But much further still to the west are the three islands of Cronos ... (and) a great mainland ... lies beyond ... at least 5,000 stader (500 miles) distant from Ogygia ... the barbarians relate ... that Cronos (Saturn) is held prisoner by Zeus (Jupiter) in one of the islands beyond Ogygia, but it would seem rather that he dwells in the great mainland that lies beyond the islands ... there, for thirty days on end, the Sun sets for little more than an hour, and for several months the night is faintly illuminated ..."

I have no idea how Plutarch came by this information, but the underlying facts must have been at least 4,000 years out of date by the time he recorded them, and they too are most likely derived from an ancient map. It was said in the past that one of the Piri Reis maps shows Greenland 'erroneously' as three islands. Yet today we know that was exactly the case until just over 5,000 years ago, because underneath the ice shields Greenland supposedly consists of three separate islands. Considering the obvious references to the northern polar region, we can now 'translate', moving in the westward direction, the names provided for us by Plutarch this way: Ogygia is the Greek name for Ireland. The three islands of Cronos are Greenland, and the great mainland is North America. The 500 miles though is roughly the distance from the southwest coast of Greenland to the coast of Labrador, not from Ireland.

Theopompus, who lived in the 4th century BCE, wrote the following: "Europe, Asia and Africa form an island surrounded by the ocean ... there is only one (other) single continent, and that lies elsewhere, it is of immense size, it breeds huge animals and men who are twice as tall as we are ..." Again, this description fits only the Americas, where the megafauna survived till about 11,000 years ago, and human giants, about 3 meters tall, (8-9 feet), until perhaps 1,000 years ago.

We could also say that fragments of the star 'maps' remained imprinted on human memory. The Druids, who used to teach about the starry heaven in fairly great detail, were also aware of mountains on the Moon, as was Democritus. Jonathan Swift wrote about Mars' two satellites 150 years before they were officially discovered in 1877. The Shilluk tribe of South Africa called Uranus "three stars." While Uranus remained unknown till 1781, it was realized later that the planet has two moons. Thus "three stars." The Dogons, of Mali, knew that Sirius was a binary star 'since ever', although its companion was not discovered until 1862. They even have always claimed this companion star was composed of a metal called sogolu, whose one grain is "as heavy as 480 donkey loads." Eventually it was calculated that a match box size of the sogolu matter would weigh one ton. It would make an exceptionally light load for all the donkeys, but the idea is there.

And that about exhausts our list of relevant examples for this chapter. Their cumulative effect hopefully proves two facts, as far as such can be proven by oral and written documents: 1) The reality of a very enigmatic past, and 2) that humans can create fairly sophisticated technology, given long enough periods of climatic and social stability, not to mention the mental dimension of each society, that would also strongly influence its focus. Our civilization, for example, is focused too much on technology, at the expense of other vital functions.

PREHISTORIC
CIVILIZATIONS

In 1997 three skulls were found along with some stone tools, about 80 kilometers from the Georgian capital, Tbilisi. They belong to the second earliest type of humans known as Homo ergaster and were estimated to be circa 1.8 million years old. The additional catch is that two of the three skulls have a cranium capacity of about 725 cubic centimeters (cc), which would put them on par with some of the smallest craniums found among contemporary, normal human beings.

A similar find was made a few years earlier, in northern Greece, located in a level estimated to be between 700,000 and 1,100,000 years old. Evidence of fire and roasted meat was also found in association with the remains.

The most striking discovery though was made in 1995 in Spain. The remains of an eleven year old child were found in strata that is more than 780,000 years old. But it is impossible to say whether by 20,000 or by 200,000 years. "What we found was a totally modern face," said the Spanish anthropologists, who eventually named this species Homo antecessor. No estimate of the cranial capacity was revealed, but since the shape of the skull's fragments follows perfectly the contours of a modern skull, the brain content would also be very similar.

But there was a 'puzzle'. The stone tools found with these remains were too primitive for such an advanced species, and more on the level of Australopitecus africanus' skills, of about one million years earlier, in Africa. But that is precisely the kind of a telltale sign we are looking for. You see, the sad fact is, archeologist and anthropologists alike are systematically making one crucial mistake: They keep ignoring fact that has been endlessly repeated since eons ago – when an advanced civilization totally collapses, its descendants are forced to restart from point zero, the most primitive level of survival. And this is also the situation with the Homo antecessor. The very combination of advanced human species and of their primitive tools point to the fact that a fairly advanced civilization collapsed about one million years ago.

The correctness of this conclusion is confirmed by the Seneca Indian 'myth' of the Seven Worlds. That is of the five previous Worlds, or cycles of civilizations, of the current one, and the one to come. An idea in agreement with virtually all major mythological systems. This particular myth tells us that the nations of the First World emerged somewhere east of the North American continent, and that, in the end, "the power of the Sun caused the devastation of the First World."

Interestingly, paleoclimatologists have only recently realized that a catastrophic warm up took place on this planet just about one million years ago. At that time, all the glaciers melted and the sea level was about 100 meters higher than today. From this we can conclude that the average global temperature most likely had reached about 25 degrees Celsius. This Jurassic type climate would have made life in the tropical regions impossible and very difficult even in the temperate regions. Indeed, the Sun, or rather the unbearable heat, along with the global flood, would have devastated the first civilization, forcing the survivors to flee northward.

The story of the Second World in fact tells us that "migrations moved toward the north." While we are told these migrants eventually created a superb culture, we are also told they followed in "the footsteps of the predecessors, who inhabited the First World. The wanton waste of the gifts of Mother Earth ... brought on ... a disease of destruction." In other words, the people kept plundering what was left of nature and its gifts, even in their reduced living space near the North Pole, which would have led to a heat death of all the life on the planet. But, as the story tells us, Swen-i-o, the Seneca's Supreme Being, "ordered the Sun to withdraw its warmth from the face of Mother Earth."

As we know by now, about 800,000 years ago a huge asteroid, of the kind astronomers call a "stone pile," exploded over the Southeast Asia, ushering in an asteroid winter, and a whole series of dreadful ice ages, making the planet even colder than it was during the last Ice Age. The longest period of stable climate, during the next 110,000 years, lasted about 1,000 years, which is hardly enough time for creating a civilization.

It is clearly symptomatic of the situation that people of that period started using sea going boats, as the evidence from Indonesia is strongly suggesting, most likely in the search for more clement regions. Anthropologists call these people Homo erectus. Yet, they are also beginning to accept the ancestors of these people have been in the region 1.8 million years ago.

In either case, conceiving of and building such boats must have required even more abstract thinking capacity than did the production of hand axes. Yet, to grasp how the hand axes work requires nowadays a university educated person and a computer.

But to continue with the story of climate woes. Since some 690,000 years ago, the climate had somewhat stabilized. At least no wholesale upheavals were reported by the paleoclimatogists, until again around 600,000 years ago. This was the period of the Third World. During this period evolved the four known races of people who created magnificent civilizations that "flourished for a longer period than the First and Second Worlds." Intriguingly, the story states this world "was inhabited by people and creatures with gifts and abilities that surpassed the two previous worlds." The 'creatures', we can quite correctly suspect, were machines, of some kind. Yet, even these people, "in spite of their knowledge … brought disruptions upon the gifts of Mother Earth." Water "was responsible for the cleansing of the Third World."

Indeed, lots of water. About 520,000 years ago, an asteroid, some ten kilometers in diameter, impacted the coastal shelf off the north-west coast of Australia. An enormous tidal wave must have washed over the continents, but mainly along the southern flank of the Asian continent, where most of the population likely lived. The consequences for the entire planet would have been horrendous.

Strangely enough, this impact had stabilized the climate. After the customary asteroid winter, the global temperature climbed from 12.5 degrees Celsius to 15 degrees Celsius, and remained there for about 50,000 years. Then, about 400,000 years ago, another massive asteroid struck in Quebec. But, from the absence of any indication of total destruction, we can only assume that asteroid impact 'merely' brought about yet another prolonged asteroid winter or, if you wish, an ice age.

Ice ages though, as I elaborate in the book *Ancient Ecologists*, are a boon for the mind, because during such periods the brain literally grows in size, and its capacity increases tremendously. The period following the impact in Quebec most likely is the Fourth World of the Seneca's myth. It became also known as the Middle

World, because, in hindsight, all the world's ideas were revolving around the inheritance this World eventually bequeathed the inhabitants of this planet.

The myth tells us the people of this period "began to keep records." That, in itself, is a rather nebulous statement that can just mean those people have invented, or reinvented, writing. More likely though, they have begun in fact to keep a record of human experience on this planet, by excavating the human past, just as we are doing, but with their mind a lot more open to the surfacing clues than ours is. Their findings must have shocked them, and made them realize their civilization was ultimately doomed as well. Doomed not so much because of the vagaries of the climate, but because of the 'uncontrollable' nature of the human beings, who tend to multiply beyond the means of environmental sustainability, especially so during the warm, interglacial periods. In the process, they not only destroyed nature, over and over again, but in turn triggered a response from the Supreme Being, or the Enfolded Order, to use the language of quantum mechanics, that can act only in the ways available to it, usually by employing an asteroid, in order to maintain life on this planet.

Most likely the group of people who were able to grasp this connection was too small to impose any meaningful changes upon the bulk of the population, and to impose them indefinitely, or perhaps the time was too short for that. In either case, they started intensely looking for an alternative solution, and they found it.

The myth tells us: "During the Fourth World, the inhabitants became aware of the universal stream that revolved around the world ..." The "universal stream" is the Milky Way. But to comprehend what the leading sentence is really telling us, we have to realize that in the northern hemisphere, we can see only one half of it and in the southern hemisphere the other half. But to the eye, even if aided by a telescope, they are both stationary. To understand that the Milky Way "revolved around the world," would have required advance astronomy, and advanced mathematics, not to speak of circumnavigating the world earlier, for the purpose of observation. In our own times we did not know the Milky Way is revolving until after the World War II.

As the relevant knowledge improved, along with the associated technology, implied even by our own development, the solution they have reached told them that the only alternative was to relocate to the space above the planet. As we shall see in the subsequent chapters, they have indeed carried out this task.

I am placing this period into the time slot of roughly 300,000 years ago for three reasons:

1. It falls between two major planetary warm ups, during which the sea level rose 26 and 32 meters, respectively, above the present level. This time slot also avoids the periods of reduced mental capacity, 'normal' for such periods.

2. Upon finding an ox rib in France, etched with Moon cycles, British archeologist Clive Gamble said: "… recent studies have demonstrated the existence, some 300,000 years ago, of mental ability equivalent to modern man."

3. Joseph Campbell estimated the age of the Bear Cult to be at least 200,000 years. In the title *Ancient Ecologists*, I explain that this 'cult' was in fact an ecological salvage organization that sprung into existence in the wake of every major ecological cataclysm. But as the indicated past suggests, such an organization could not have been always so readily available, had it not have a permanent location somewhere off this world.

We are also told that "the greatest span of existence was experienced by the people evolving in the Fourth World," who populated the entire planet. While it is possible that the planet-wide civilization might have endured for tens of thousands of years, i.e. for the duration of the ice age, the leading elements of this civilization, and thus, in a certain sense the civilization itself, has endured for some 300,000 years, albeit for most of this time in the space habitats above the planet.

"The duration of the Fifth World was short, compared to the previous worlds," the Seneca legend informs us. From this we could deduce nothing of note had happened in the human affairs until way after the next major global flood that took place about 100,000 years ago. The sea level then rose 30 meters, above the present level, from an unknown depth. At the same time though, this climatic swing was so severe that current estimates claim only some 10,000 human beings have survived that cataclysm. The severity of conditions during that period is further confirmed by yet another statement from the Seneca source: "Man has passed through many environmental experiences. His lessons were extremely difficult …"

Indeed, after the initial warm up, the glaciers started to grow again, some 75,000 years ago, after an eruption of a mega-volcano in Sumatra, compared to an impact of a 10 kilometer asteroid. Then, 64,000 years ago, it got warmer, for a change. Yet, only 4,000 years later, maximum glaciation occurred, and the Arctic ice shield reached the Baltic Sea. Great climatic oscillations took place between 37,000 and 30,000 years ago. About 20,000 years ago another maximum glaciations took place. This time it got so bad, the sea was some 120 meters below the present level, glaciers filled the Death Valley, and Amazonia was a cold savannah. This particularly severe ice age ended suddenly, in a great cataclysm, some 15,500 years ago.

The Seneca legend also informs us that "the records of the ages," left behind by the people of the Fourth World, "were being uncovered, and the false documents of man were being corrected." Here the word "documents" is a synonym for the word "records," which we have identified with science. What we are being told in fact is that the flawed scientific assumptions of the times have been corrected by a discovery of the (then) ancient knowledge.

And still, "man was having difficulty practicing the decree of the Great Spirit ... (as) wars within his attitudes and thoughts still festered in his mind, creating injustices upon the gifts of Mother Earth." And great, terrible injustices they were, as we shall soon see. "The cleansing (this time) was exerted by the powers of the Sun and Moon ..." This 'Sun' though refers to the heat of the nuclear holocaust that burned the Earth. The Moon refers to the subsequent reduction of mental powers.

The Sixth World is not a world of a past civilization, but the very world we live in, the world as it had existed since the cataclysm that ended the Ice Age, about 15,500 years ago. Seneca Indians, like most ancient cultures, maintain the tradition that this is the last stage of our guided evolution, the last period during which mankind is to learn some harsh lessons from its own stupidity.

"For all six worlds, man has wreaked havoc upon himself, his fellow man, and the creatures of nature. Now he (stands) at the threshold of perfection, awaiting the wisdom of the ages to penetrate his mind."

Although the legend speaks of it in the past tense, the cleansing of our world is to be exerted "by the power of the Moon, followed by the heating properties of the Water ..." The "power of the

Moon," as noted above, refers to man's reduced mental abilities. This reduction started to take effect since about 11,700 years ago, when the human cranium, and the brains within it, started to diminish in size and performance – as it obviously does during each and every interglacial period.

The "heating properties of the Water" is merely an indirect reference to the greenhouse effect. After all, to warm up water requires heat. Since it was the water in the natural setting that was referred to, a substantial increase in atmospheric temperature was obviously alluded to.

Of the Seventh World the Seneca myth tells us only this: "The final cleansing had been completed, and man's life (is being) guided by ... the same (spiritual) light that is in the mind of Swen-i-o, the Great Spirit."

Summing Up

In the mythologized annals of the human past, the Seneca account has no equal among the cultures of this world in its detailed accuracy. While most major cultures have attempted to remember, one way or the other, the preceding cycles of civilizations, only the Hopis, and the Mayans, at best hint at the possible causes of their demise – and most likely because they have shared in a common heritage, upon the North American continent. It is also only these people, along with the Soninke people of Africa, who not only remember the past, but also predict the future.

On the other hand, while the Persian myths, as far as I know, do not provide any such overview of the past civilizations, their rendition of the future changes is more accurate than that of the Seneca. As to the additional causes of the final cleansing, the Jewish 'myths', contained in the Bible, are more accurate still. But Nostradamus is even more accurate; in my interpretation, anyhow.

Of course, our immediate concern here is not the future, but the past, and more specifically still, the incredible period that started about 300,000 years ago.

UP, UP, BUT NOT AWAY

Reading the Seneca 'legend', we cannot avoid realizing how extremely difficult it is to survive the cataclysmic events associated with the major climatic swings. Furthermore, since most of them were caused by ill-considered human activities that resulted in wrecking the life-supporting ecological systems, which the people of the Fourth World would have found out about through their archeological and paleoclimatic research, one could say the difficulty to maintain an advanced civilization was, and is, directly linked to the human 'lack of evolutionary advancement', or just to plain old stupidity, and inadequate imagination.

It must have dawned on the brightest and the best organized portion of the population, sometimes between 300,000 and 400,000 years ago, that the safest bet (at the time) was to part not only with the bulk of mankind, but with the Earth as well, in order to avoid the consequences of mankind's activities on the planet. The only venue of escape left to them was the space above the planet, and they took it.

This very motivation was captured in 'myth' as well. In the conclusion of a story describing a mutual slaughter of the supporters of Ra and of his enemies, Ra announced: "Henceforth my dwelling place must be in the heavens." Then, according to one of the Egyptian creation myths, Ra "called into existence the field of Aalu, and there he caused to assemble a multitude of beings ..."

Of course, nothing is as simple as it sounds in any single myth. There never was any magic, and everything people have achieved, they had to work for pretty hard. Thus, in the attempt to 'liberate' themselves, the people of the eons past have made their first forays into the space and assembled their construction platform there.

Ra called unto Thoth, the God of Learning and Wisdom, and announced: "I will make something shining and resplendent in Duat (Other World) and the Land of the Deep." (As we shall see, the space above the North and the South Poles were meant.) "I cause thee to embrace the two heavens," commanded Ra, and continued with "I cause thee to turn to ... the North" – thereupon the Cynocephalus (i.e. the "Doghead") of Thoth ... came into being."

(The meaning of the "Doghead" shall be explained shortly.) At about the same time the sacred 'bird', Tekhni of Thoth, came into being. (Again the mythological reversal, as the 'bird' would have come before the 'Doghead'.)

Virtually the same ideas are expressed in the Mayans' sacred book, the *Popol Vuh*, where the character named Seven Macaw represents also the early stage of establishing permanent habitats in the space above Earth: "I am great. My place is now higher than the human work … since my chest is a metal, it lights up the face of the Earth. When I come forth before my nest, I am like the Sun and Moon … my face reaches into the distance." (I.e. when in view, the space station shines brighter than a star.)

Zipacna, the Seven Macaw's 'first born son', is credited with building the six great mountains. His 'younger son' is named Earthquake, and it is stated that "the mountains are moved by him." Well … the name Earthquake, or Cabracan, in the classical Quiché language, can also be translated as "leg," "trunk," or "pillar." As we shall shortly see in greater detail, Cabracan indeed represented that, which in our age was first thought of, over 100 years ago, by the Russian scientist Konstantin Tsiolkovsky, who named it the space elevator. More recently, scientist and science fiction author, Arthur C. Clark, proposed that such an artifact, made out of super-light and super-strong materials, would be 'suspended' between the space and the planet, providing an easy and economical, as well as ecologically safe access to the space from the planet, and vice versa.

In this context, Zipacna would represent the space platform, assembled in space with the help of a sacred 'bird' (i.e. a space shuttle), and this platform first served to construct the Cabracan, i.e. the Space Elevator, at or near the North Pole. The name "Doghead," or "Dog Star," was originally assigned to the space elevator, before it was later transferred to the constellation we now call Sirius. The reason for that is not exactly clear, but most likely is connected to the ancient notion that dogs are the guardians of secrets. It was literally through the Cabracan that certain components of the future 'mountains' were moved then into the space. In this fashion he then 'moved the mountains'.

The 'mountains,' of course, are the space habitats themselves, known also as the revolving castles, among the Celtic people, as a firmament among the Jews and the ancient Finns, or as the Sky Homes of the Gods, Spheres of Heaven, or just Nests, elsewhere.

Since the people on Earth perceived the seven structures (six space habitats, plus the space elevator) as seven stars, they also referred to them variously as the Seven Rulers of the World, Seven Mothers of the World, Seven Hathors, Islands of the Blessed, Seven Sages, etc. The Jews, unaware, still reflect this grouping in their menorah – a candelabra for seven candles, with the stand representing the space elevator.

The very name "firmament" suggests something solid. And indeed, not too surprisingly, the Hebrew word for firmament, Rachnia, is derived from the word for beaten metal, and the Germans translate this expression as "ein veste," which can mean a citadel, a fort, as well as a firmament. Furthermore, it is written quite plainly in Genesis that "Yahweh made the firmament." (I:7) Illmarinen, the Eternal Smith of the Finnish Kalevala, also stated: "I forged the firmament."

E.A. Wallis Budge informs us that "the primitive Egyptians believed that the floor of Heaven … was made of immense plate of iron, and that it rested on flour pillars, or pylons." Interestingly enough, the Egyptian divine "blacksmith," Ptah, was also recognized as the "Father of Gods and Men." The Sumerian 'Father' of the Gods, who created the Heaven (that is the firmament), is named An, which, in Akkadian, meant metal. The natives of the New Hebrides also used to believe the sky was a rock, (i.e. metal), and that in some places it was rooted in the land.

In a certain sense Homer and Hesiod continue with the imagery of metal, and picture the world in the shape of a round shield. Homer finishes the description of it: "And thereupon Hephaestus set up the great strength of River Ocean, beside the outermost rim of the shield …" (*Illiad*) While the contemporary critics argue such a River Ocean lacks a clear concept of another bank, they fail to ask what had the divine blacksmith to do with "setting up a river." The answer to that partially rests (again) with the meaning, the original meaning, of the words used. The expression River Ocean is derived from the Sanskrit expression "a-Cayana" or that "which encircles." Here namely the planet Earth, being encircled by the atmosphere, which reaches about 80 kilometers (50 miles) into the space. Obviously, this 'atmospheric stream' has no other 'bank' other than Earth – until an artificial one is created somewhere at the opposite 'side' of such a stream, which is what the divine smiths did.

In a Sumerian myth I came upon a remark about "a man of the Nether World River," referring to the same concept. Apsu, in the Babylonian myths, was the domain of 'sweet water', located 'beneath' the Earth, home of Ea, the God of Wisdom, and of the Seven Sages. The Yahweh also ordered: "Let there be a firmament in the midst of waters." (That is the atmosphere.) Obviously, since the space habitats could not have circled 40 kilometers above the surface of the planet, the Space Elevator was somehow vaguely referred to as 'dividing the water'. But most likely only a fuzzy memory of "something" was being voiced.

Both Homer and Hesiod, when referring to this (for them) nebulous region, employed the concept of "peirata gaiés," or "the borders of the Earth." This concept was also hotly debated about 2,000 years ago by the Alexandrian scholars, who agreed that Homer had situated the Island of Calypso in "indefinite and *outward* removed places." Still, Homer also described the shoreline of Polyphemus Island as "peirata gaiés." Hesiod transferred the same name to the 'underworld'.

Here, perhaps, might be a good place to mention that such lore even used to remember the distance of the Revolving Castles from Earth. Several peoples, including the Egyptians, made an attempt to 'peg' the distance, but it was the Chinese who got the closest, on at least one occasion. This particular one mentions that the distance from Earth to Heaven is 510,000 li, which equals roughly 280,000 kilometers, or 170,000 miles, and that in turn fairly closely corresponds to the Lagrangian points L1 and L2, where the Revolving Castles would have had to have been located.

While most astronomers do not really know where exactly the Langrangian points are located, as one of them told me, these are usually calculated as being situated between 64,600 and 58,000 kilometers from the Moon, towards the Earth. At the same time the distance of the Moon from the Earth ranges from 357,000 to 408,000 kilometers. By subtracting the lesser figures from the greater figures provides us with the Langrangian points. The L1 is located 'above' the North Pole and the L2 is located 'above' the South Pole. (Re: The 'Underworld'.)

We also have to mention another technical element of construction of these Revolving Castles, namely the use of solar panels and/or light absorption and transformation panels. In the Bible we can read that God "spread out the skies, hard as molten mirror." (Job 37:18) The Australian aborigines claim that the vault of Heaven was made of rock crystal, the "stones of light" to the Dayaks of Sarawak. In the Ethiopic *Book of Enoch* we can read that he was elevated to heaven, till he "arrived at a wall built of stones of crystal." Inside was "a spacious habitation, built also with stones of crystal. Its walls too, as well as pavement ... its roof had the appearance of the running of stars."

Having the light 'crystals' everywhere suitable would suggest energy was being recycled to the utmost. The observation about the roof and "the running of stars" was also recorded in the Slavonic legend of St. Catharine, and it speaks quite plainly about the speed of rotation of these structures.

Now let us return to (the idea of) the space elevator, to see just by how many people it was remembered by, and the various ways in which it was being described. The most simplistic one that comes to mind is to be found in the story of Jack and the Beanstalk. The Navajo Indians picture it as a fast growing hollow reed, planted on a mountain. The Hopi Indians talk of it as a fast growing bamboo. The Polynesians chose a strong vine tendril. Four different tribes in Africa think of it respectively as trees, poles, porridge pestles, or masts on the top of each other. One of the more popular images was a ladder often mentioned in the Egyptian Pyramid Texts, which also credit Ra with making it, while also referring to it as the ladder of Seth. It is known as such in stories throughout Africa, where it was often portrayed in the rock paintings. This name was also fairly common in China and Japan. Even the Bible mentions one – the Jacob's ladder.

A slightly more sophisticated, but confused concept reaches us from Malaysia. It is about a blacksmith Budar Yoid Intoie and his "magic knife." He made it long enough to bridge the sea, but it somehow ended up "faraway above him in the sky." The Betsimisaraka people of Madagascar 'echo' this image by offering us the idea of a silver cable. The Persians believed a Kinvad bridge connected Earth to the cosmic mountain. In the Nordic myth it is called the Bifrost Bridge, "fragile and steeply poised above the abyss ..."

A giant tree connecting the Earth with the sky was also a favourite image. In Mesopotamia it was variously called The World Tree, The Black Tree Kishkan, or The Chulupu Tree, spreading over an ocean. There was also a World Tree known as Mésu, whose name is most likely related to the Mashu Mountain. (See below). This Mésu tree was said to have 'roots' a hundred miles long, or some 160 kilometers. In Scandinavia they referred to it as Yggdrasil Tree, and conspicuously mention it was supported by three mighty roots.

Another favourite image for the space elevator was that of a mountain. Herodotus describes the Atlas Mountain, upon which was based the myth of the giant Atlas, supporting the sky, this way: "It is very narrow and a complete circle. It is said to be so high that one cannot see the peaks of it, for summer and winter, the clouds never quit it." (4.184) Gilgamesh, the semi-divine hero of Mesopotamian myths, had traveled over twelve leagues (about sixty kilometers, or fourty miles) through such a 'mountain', or rather what was left of it. For the Hindus it was Mount Meru, that stood directly under the Polestar. That is 'right' at the North Pole. Interestingly enough, the people who used to reside in Siberia, such as the Lapps, Finns, Estonians, and the Samoyed tribesmen, who still do, call the North Pole the Sky Nail. Buryat, Kalmyk, Kirghiz, Bashkir and Innuit, all circumpolar people, agree that some sort of pole, or nail, or pillar, connected the Earth to the Polestar. The Teleut refer to is as the Sun Pillar.

And 'pillar' was absolutely the most favourite description of the Space Elevator. The Egyptians mention the four pillars of gods fairly often, without being specific about them. So do the Chinese, who mention their supports, in form of 'cords', as well. The Mayans, it you remember, also opted for this image. And so did the Finns. In the Kalevala it is called the Sampo, and its three roots were said to be each nine fathoms deep, which is 54 feet, or 18 meters, but that most certainly is an irrelevant measurement. It was "Illmarinen, eternal smith ... (who) forged the Sampo ..." The name Sampo itself presented a bit of a puzzle for the translator, as it 'seems' connected with the word "sammas" (genetive sampaan), which means a pillar, or a post. (A tentative evidence for the three support structures can also be found in the linguistic and astronomical analysis of certain Hebrew words found in the Old Testament, but it is too convoluted to be introduced here.)

If we were to partially summarize what we found out about the space elevator so far, we can safely say it was an immense structure. The current proposals envisage a structure ranging from 101,000 to 136,000 kilometers in total length, floating in space, barely touching the Earth. Even if very slender, it would have been at least two to three kilometers in diameter, and constructed out of (semi) flexible sections. It would have been 'glossy', or metallic in appearance, and anchored to Earth's surface by three 'cables' of an equally awesome size.

According to the tradition, one of these supports would have been anchored in the Atlas Mountains region of northwest Africa, one somewhere in northwest China, and one in southwestern North America. If so, these 'roots' would have been neither 18 meters, nor 16 kilometers long, but staggering 4,500 kilometers long, or roughly 3,000 miles – without allowing for the curvature of the Earth. One does not have to be an engineer to realize such 'cables' would have collapsed under their own weight, and rip the space elevator apart. Obviously, placing such structures in those localities was nothing but vague memory, combined with fantasy and wishful thinking. In reality these 'cables' could not have been more than several kilometers or tens of kilometers in length.

The Celtic story known as "The Voyage of Mael Duín" most likely describes an actual encounter with a remnant of such a support 'cable'. "… they came upon a great silver pillar with four sides, each side being two oar strokes of a boat in length ... and there was not a single sod of land round it, but only the boundless ocean, and they did not see the nature of its base below, nor of its apex to the top. There was a silver net reaching far out from its top and the boat came under sail through the mesh of the net; and Díurán gave a blow with the edge of his spear across the mesh of that net. 'Do not spoil the net', said Mael Duín, 'for what we see is the work of mighty men.' "

Obviously, this report differs from the one provided by Hesiod who *heard* of a "very narrow" and circular structure, while the Celtic account speaks of a four-sided structure, of no more than several meters on each side. But both versions could theoretically be correct, because, as we shall see later, the Space Elevator, and most likely the support structures as well, were repaired, or replaced, at least once or twice, within the last 20,000 odd years.

The Mael Duín's party later "saw another island, on one leg … and rowed around it, looking for a way into it, but did not find any … but below the surface, at the bottom of the leg they saw a locked door … but they did not speak to anyone and no one spoke to them. They retreated …" Although the story mentions no further details about the size of this 'island', it would seem it was rather large, and its designation, "on one leg" echoes the Mayan name, Cabracan, by which the Space Elevator was meant, and that also was translated as a "leg," or "pillar."

Most likely, by then, the 'leg' was ruined and abandoned, or they were simply uninvited guests. Those with the visiting privileges first had to be brought to the Earth side terminus of the space elevator, which door was indeed located below the surface of the sea. An Innuit story, from the Bering Strait, speaks of a man who was taken by a 'raven' to the Skyland, where it was very beautiful, and the climate was much better. They flew back down "through a star hole, floated a long time through the air, then through something else. They were at the bottom of the sea." A Hopi story tells us of two 'birds' who "flew to the lower world ... and back" through the Sky Hole – together.

Such stories not only tell us about the undersea terminal, but they also make it clear the space elevator was spacious enough to accommodate crafts that could fly in the space and atmosphere, and transport them, somehow, up and down, through its interior. Presumably such crafts were also able to navigate underwater, just like the UFO's.

The Hopis have preserved for us possibly the most technically satisfactory description of gaining access into the Space Elevator. Yet this one too is flawed. Living in an arid region of the continent, they substituted a lake for the Arctic Ocean, in the story named "The K'Yaklu Being and the Duck."

"... they lightly bore him (K'Yaklu) to the shores of the deep black lake, where gleamed from the middle the lights of the gods ... and bore him over the water to the magic ladder of rushes and canes which reared itself high out of the water. And K'Yaklu ... stepped down the way, slowly descending a sky-hole. No sooner had he taken four steps than the ladder lowered into the deep ..."

The story tells us, in an awkward, confused fashion, that the party had somehow reached the space elevator, at the side of which reared up a commodious 'elevator cabin', and with the passenger aboard, descended towards the terminal. Needless to add, the submersible cabin would have been pressurized and watertight like a regular submarine. The story also makes it clear that any entrance into the space elevator was called a "sky-hole," and did not refer to any "hole" in the sky.

For example, in another Hopi tale, "The War Chiefs Abandon Iatiku" we read the following: "Horned Toad opened wide its mouth and Humming Bird flew in and right through him, down the (sky) hole." In another story we read: "... the old Centipede wriggled himself up the sides and over the roof, and down into the Great sky-hole."

These images though bring us to the space side of the space elevator, describing the activities there. Sort of. The "Horned Toad" probably represents one of the three (or alternately four) pressurized lock gates, allowing various space-going vehicles to enter the space elevator, then to be conveyed somehow to the Earth side terminal, and be able to fly from there all over the planet. The Centipede possibly represents a maintenance robot, of some kind. And although one has to be careful with such a designation, it seems a good possibility here, as we shall meet more robots later.

While at the space side of the structure, we should point out the 'myths' also explain its construction started in space and was gradually extended towards Earth, as modern engineers envision such a job should be done. The Mayans have preserved for us the memory of the original space platform, where all the work begun and the Japanese preserved the following key information, in their Creation Myth: "... a certain thing was produced between Heaven and Earth. It was in form like a reed-shoot. Now this became transformed into a Mid-Sky-Master, and was called Land-eternal-stand-of-august-thing ... it had no place of attachment for its root." Then the story makes a mental somersault and tells us that the first three stages of this "reed-shoot" became deities who "thrust down the jewel-spear of Heaven," (literally a male-pillar), "and groping about therewith found the ocean. The brine which dripped from the point of the spear coagulated ... so they made Spontaneous-island, or Ahaji, the pillar of the center of the land." The very next sentence states: "Two Kami (gods) stood on the floating bridge of Heaven ... saying: 'is there not a country beneath?' "

Incidentally, contemporary ideas also call for such a floating bridge, projected to be 36,000 kilometers above the planet, to be constructed first, beyond which the elevator would project further 110,000 kilometers into the space, as a counterweight.

While ignoring the understandable confusion, we can quite clearly see that first were constructed the three topmost sections, recognizable even on the Egyptian Tet 'tree', as its three, slightly

The illustration above represents one of the four best known fetishes from ancient Egypt, known as the Tet column, or Tet tree. It most likely expresses memory of the sectional space elevator, recounted in so many myths, the same way the Jewish seven-branched menorah 'remembers' the seven lights people used to see in heaven.

conical sections, joined by four discs. Next, the pillar was extended all the way to the Earth's surface, where it was anchored, ('rooted'), to keep it stationary there. The bulk of this structure floated 'above' the planet, and above the floating bridge, in order to avoid being crushed by its own weight.

We can even deduce that the structure that used to connect the twin worlds, the one on Earth and the one in space, was nicknamed 'twin,' judging not only from the reference to the twin Kami gods 'standing' there, but also from the reference in the *Epic of Gilgamesh*, where the 'mountain' at the edge of the world, and leading up to "the sky's foundations," through which the hero traveled, is named Mashu, which also means a 'twin'.

Constructing such a 'twin' must have been an awesome undertaking that would not have even been considered by any visiting aliens. The almost ever-present verbal testimony of its existence proves two things: 1. It was placed there by our long ago ancestors, who then proceeded to construct the six artificial habitats, further away from the planet. 2. The existence of the space elevator and of the space habitats also demonstrates their intention to 'remain in orbit' in perpetuity.

REVOLVING CASTLES

"To Him that rides upon Heaven of Heavens, which were of old ..."
Psalm LXVIII:33

"There lies a land outside the Heavens, outside the pathways of day and of the Sun, where Atlas, the sky-bearer turns on his shoulder the wheel inlaid with bright stars." Virgil in *Aeneid*

As we have already noted, *Popol Vuh* informs us, Zapachna, the Seven Macaw's son, is said to have built the six mountains. It also tells us that "the mountains were separated from the water, all at once the great mountains came forth." Thus far the mythologized part of the event. Yet, the very next sentence is pretty plain, straightforward, speaking obviously of the ancestors who did the work: "By their genius alone, by their cutting edge alone, they carried out the conception of the mountain-plain, whose face grew instant groves of cypress and pine."

Once properly understood, the above represents the shortest and the most breathtaking sentence, describing the most profound leap in the evolutionary history of mankind, so far. At the same time, it also manages to convey the main visual feature of the space habitats: the inverted world inside an immense hollow ring, whose habitable part would give the impression of an odd valley, with relatively flat bottom, and steep sides, that most likely were shaped like mountains, and planted with trees. Thus "the mountain-plain." Several myths also provide the valley with a river. The ancient Egyptians, as if to confirm, used to say that the Nile Valley resembles the home of the gods.

We also ought to notice there were six such satellites constructed, and that the "separation from the water" is referring to the substantial distance from the Ocean Stream, the moisture laden atmosphere.

In the so called creation myth from Tibet, we read the following account: "Formally by the magical power of the gods, the gSas and the dBal, an egg formed of five precious gems burst open by its own innate force from the celestial womb of the empty sky. The shell became a protective armour. The tegument defending weapons, the white became a strength-potion for heroes, the inner tegument became a citadel for them to dwell, that obscure citadel, the 'Sky-Fort of the Waters of Wrath,' it stole the Sun's light, so bright was it. From the very inner part of the egg there came a man of magical powers …"

The above is a typical example of a 'marriage' between a few crumbs of facts, and an impoverished flair for poetry. And yet, as so often happened, this strange bond preserved a few additional key facts: 1. The outer shell of the space habitats was hardened against cosmic radiation, the protective armour. 2. It must have also bristled with 'weapons' meant to protect it from small asteroids and meteors. That these were energy weapons is made clear in another episode from the Mael Duín's travel, when the voyagers find themselves upon a small island, whose white ramparts "almost reached the heaven." One of them stole a huge silver and gold 'collar' and "the (guardian) came after him, and jumped through him like fiery arrow, and burned him to ashes."

At the same time, the Tibetan account repeats the idea of 'water,' named here the Waters of Wrath. (Later it shall become obvious why they acquired such an ominous name.) And in contrast to the "magical power of gods," we also have here "a man of magical powers." No doubt they were the same. That the 'gods' of the ancient times were people of this planet is confirmed from many diverse sources.

The Seneca Indians say: "A long time ago, human beings lived high up in what is now called Heaven." The Greek poet Pindar stated: "There is a race of men, there is a race of gods. Both drew the breath of life from the same mother, but the powers are separated, so that men are nothing, and the others are masters of the luminous heavens, which are their citadel for ever." Even the Australian aborigines, who connect most of their lore to the nebulous "dream time," which is really the present, also maintain a memory of "long ago, back beyond even the dreamtime, there was a group of people who did not live on Earth. They lived in the sky world."

In the Gnostic texts found in Egypt, we can read: "Seven appeared in chaos," i.e. the sky. Yaldabaos, "… he created heavens for his offsprings … created them as dwelling places … and they were completed from his heaven to as far up as the sixth heaven."

The text also explains to us this 'chaos' in fact "is a kind of darkness and this darkness came from a shadow," and that this shadow in turn "comes from a product that has existed since the beginning." Essentially we were told the 'heavens', or firmaments, were created in the darkness of space, and as a 'bonus' we are also told, in a convoluted, 'esoteric' fashion, how the space itself came into being. The text even tells us the name Yaldabaoth meant "child, pass through to here." Such a name again suggests a conveyance through something; in our case through the space elevator. The six 'heavens' of his 'children' are the space habitats, which idea also confirms the sequence of events, as did the Mayan version. Uniquely, the text even mentions a heavenly administration.

The paradise of Hindu gods was also perceived as seven separate cities and palaces, amidst green woods and murmuring streams, situated in seven 'circles', placed one above another. Of course, the 'above' depended upon one's perspective. In space, three and three were 'above' each other, and in turn the six of them were above the space elevator.

Incidentally, the world Paradise is derived from the Persian word Pairidaeza, which most likely was related to the Persae Tree, and that was yet another name for the space elevator. And by the same token, the Greek expression "peirata gaiés" also evolved from this name. In the Indo-European language the root-word "per" formed a base of prepositions and preverbs with the basic meaning of "forward," "through," "toward," etc., all signifying a traffic of some kind, and the direction of the same, and the word "daesa" used to mean a "wall" in ancient Persian. On the other hand, the Persians have also believed it was the good god Ahura Mazda who created the Universe, while Ahriman, or Aharman, the 'evil' god, "produced those seven planets," which is how these seven structures had appeared from Earth.

To generate an artificial gravity, the space habitats would have to be circular, and revolve fairly briskly around their axis. At least that is what the contemporary concepts call for. The Revolving Castles, as these structures were known to Celtic people, were designed on the same principle, and its basic design was preserved for us in the form of a Ferris wheel. No, not the mainstay of amusement parks invented by Mr. Ferris, but as described centuries ago, although the name applied to it back then was spelled fairies wheel. And the said fairies, of course, were the aerial race, living in the so called "sidhs," of which we shall talk a lot more, shortly. These are closely related to the Wheel of Arianrhod, also of the Celtic lore, of which we shall talk quite a bit as well.

If it needs pointing out, where the current Ferris wheel is rotating vertically, the fairies wheels rotated 'horizontally'. And they were much, much larger. While our far removed ancestors were able to construct the space elevator, which was some 36,000 kilometers long, not counting the 110,000 kilometers long 'counterweight', the Revolving Castles would have been constructed on a more 'humble' scale, with a circumference of perhaps around 2,000 kilometers. For example, the dimension of the New Jerusalem, as given in *The Revelation*, is to be about 12,000 furlogs, or about 2,400 kilometers, reflecting most likely a memory of the space habitats. To maintain the rigidity of such a structure, numerous supports, or 'spokes,' radiated from the hub to the encircling 'wheel' of the space habitat itself.

Unfortunately, I was not able to cull any other specific reference to the size of any of the Revolving Castles, but whatever their real proportions were, they obviously overwhelmed the visitors from Earth. At least judging from the several such references we have.

In the Indian epic *Mahabharata*, in section known as Arjuna's Journey, we can read: "When he approached the region, which a mortal walking on Earth never sees, he saw thousands of wondrous vehicles ... there neither the Sun, nor the Moon shine, not even fire burns there. That, which to the people on Earth below seems to be stars, distant lamps, in fact are immense structures."

This account, which should be filed in your memory for future reference, is as straightforward and 'modern' as one can ask for and, in a sense, self explanatory, because the Sun and the Moon shine on Earth, and the fire burns here, only thanks to the atmosphere. It reflects the photons, it provides oxygen for the fire to burn. Yet, these facts were most likely unknown to the people who recorded the epic, some 3,000 years ago. Only tradition kept these tidbits in circulation, including those in the following Hopi story.

"At last they came to the House of Beloved Gods themselves. The red house was a wondrous terrace, rising wall after wall, and step after step, like a high mountain, grand, stately. But the walls were so steep and smooth and high that the skill and power of the little War Gods were of no use."

This account tells us this particular space habitat, constructed from a red metal, was shaped like a (round) stepped pyramid, and high as a mountain. Since a stately mountain ought to be at least 2,000 to 3,000 meters tall, (6,000 to 10,000 feet), and since this 'mountain' had its "up" and "down" side, we are talking of a very respectable girth here. Considering these spectators were describing only the segment of structure immediately facing them, again, no more than a few kilometers on either side, the total circumference must have been awesome. And to underline the obvious, this particular structure was not of the 'simple wheel' type, but a tiered model, shaped like a 'spinning top,' or possibly of a spiral design. It is likely this structure was rendered in a Chinese woodcut. (See the next page.)

Aenas, the fabled semi-divine king of Troy, after visiting the 'Underworld', as did all the sacred kings, commented on what he saw there: "… the battlements were forged in the furnaces of the Cyclopes," which immediately suggests 'cyclopean' dimensions. This can be confirmed from the Scandinavian mythology, where it is stated that "the Gates of Utgaard," the home of Scandinavian gods, "were so high, one had to bend backward to see the top," or somewhat like trying to see the top of a skyscraper, while standing in front of it. And this reference is only to the gates, not to the rest of the structure.

We can also glean a few basic facts about the materials used to construct the space habitats, either from the lore, as well as from the meteorites recovered. To start with, no ordinary metal could have been used to construct the space habitats, because at the Lagrange points the temperature is approaching absolute zero; that is -273°C, or -460°F. Even if 'heated' from the inside, on the outside the structures still would be very cold. Ordinary steel, without nickel, becomes brittle at -30°C. With only 0.4% of nickel it can withstand -40°C. With 5% nickel it can already withstand -60°C.

Now a bit of a 'puzzle': The vast majority of meteors, or 94%, are supposed to be stone meteors, and some of them, like olivines and chondrites, also contain iron, as does the Earth crust. Yet, 35% of all meteorites recovered are 'iron' meteorites and they typically consist of 90.8% iron, 8.5% nickel, 0.5% cobalt, 0.17% phosphorus,

The reproduction of the ancient Chinese woodcut, above, portrays a couple of water dragons, ready to swallow one of the space habitats, such as the one mentioned on the preceding page.. Yet, while it clearly portrays the rising waves, centre-bottom, induced most likely by the front of the immense pressure wave on the sea surface, caused by the structure, the habitat portrayed here is almost certainly of the tiered variety that crashed in the Sahara.

0.04% sulphur, 0.03% carbon, 0.02% copper, and 0.01% chromium. The content of nickel ranges from a minimum of 5% to a maximum of 34%! Such alloys would seem perfectly suitable for the construction of space habitats, and are noticeably different in composition from the near Earth asteroids, or the Earth crust. The asteroids contain 37% oxygen, 26% iron, 18% silicone, 14% magnesium, 1.5% aluminum, 1.3% calcium, and only 1.4% nickel. The Earth's crust contains about 50% oxygen, 25.8% silicone, 7.5% aluminum, 4.7% iron, 3.4% calcium, and 2.6% sodium. It is no wonder then that the meteoric 'iron' was so sought after in the ancient times. It was, in fact, a top quality steel.

The "Red House" of the Hopis is most likely the same as the "Hephaestus Mansion" in the sky, or the "copper walls of Tartarus," in the 'Underworld'. However, this 'copper' is unlike any copper we know of. It is said that the Inca noblemen used copper swords that were better than steel swords. The mythological Yellow Emperor of China also made himself a "sword from the red copper of the Kunwu Mountain. It was green in colour and cut iron and jade as if they were but lumps of soft Earth," recorded the Chinese annals.

Without a doubt, the people living inside the space habitats would also like to see a blue sky, during their day, up there. This effect was most likely provided by a crystalline structure resembling the famous lapis lazuli, described as an opaque, azure-blue to deep blue gemstone of lazurite. Yet, when the Bull of Heaven fell to Earth, Gilgamesh called the blacksmith to "dismember it" and to remove the lapis lazuli it was covered in. (At least that is my interpretation of this fragmented, convoluted Mesopotamian story that offers only hints and suggestions of the real events.) A Glass Castle is also known from Celtic lore, where the word "glass" is said to have meant "the blue of Heaven," in Gaelic.

Before we embark on the last section of this chapter, we should review the technology associated with the Revolving Castles. We have already talked about the genetic lab and biological suites, the energy weapon(s), as well as the spaceships. One of them, mentioned in the Egyptian Pyramid Texts, was about 2,000 feet long, or over 600 meters, which is roughly the size of Voyager, the spacecraft from the television series Star Trek – The New Generation.

The ancient lore and myths also speak of the epitome of any futuristic fantasy: the robots. We have already described the Hephaestus' 'Golden Girls,' and his 'cauldrons'. He is also known to have constructed Talos, a mechanical giant who is said to have later protected Crete. Another such 'character,' Humbaba, from the *Epic of Gilgamesh*, used to protect the huge Lebanese forest, once the home of the 'gods'. We have also mentioned the story from China, about the man name Li, who had eight silver 'men' in his treasury, who walked out one day, turning into young men who could talk. Just like the Hephaestus' 'Golden Girls.'

But we have saved the most telling story for the last. It comes from the New Hebrides Creation Myth. The themes are familiar: Naareau the Elder "made that thing in the Void," and Naareau the Younger got stuck on the planet. He "walked on the overside of the sky that lay on the land. The sky was rock (i.e. metal), and in some places it was rooted in the land ... he said, 'I will enter beneath it.' " He voiced a spell, we are told, and gained access. He knelt on the sky and began to tap it with his fingers, saying:

"Tap ... tap, on Heaven and its dwelling places.
It is stone. What becomes of it? It echoes! ...
Open Sir Stone! Open Sir Rock!
It is open-o-o-o!"

Then "he looked down into the hollow place. It was black dark, and his ears heard the noise of breathing." The first creature that came out was a bat. Since a bat presumably can 'see' in the dark, it informed him: "I see people lying in this place ... (but) they move not, they say no word ..." Naareau the Younger assumed they were "Fools and Deaf Mutes ... a breed of Slaves. Tell me their names," he asked the bat. The creature answered: "This man is Uka the Blower. Here lies Naabawe the Sweeper. Behold! Karitoro the Roller-up. Now Koteka – teka the Sitter. Kotei the Stander now – a great Multitude."

He went inside and tried talking to them, with no result. Then he shouted to them "Move!" and they moved. Soon they started to work and to communicate with him.

What this story is attempting to impart is about the existence of specialized robots, 'asleep', until ordered to perform their tasks, such as sweeping the grounds, babysitting – or supervising, repairing solar shutters, etc. Since there was a great multitude of them, for every conceivable type of repetitive work, manual or 'intellectual', (which does not require an original input), the sky people certainly did not have to work too hard – if at all.

And finally: Some years ago a group of Siberian fishermen hauled in a piston, made of an alloy containing 67.2% cerum, 10.9% lanthan, 8.78% neodyme, and traces of iron, uranium and molybden. It was made out of a metallic powder, under pressure hundreds of thousands times greater than the regular atmospheric pressure, a pressure which our technology is still unable to achieve. Furthermore, it was made in space, about 100,000 years ago, concluded the (then) Soviet scientists.

The Persians used to believe Ahura Mazda created the universe and Ahriman 'only' the seven planets, and these, it was later assumed, were the planets of our solar system, which in turn made some people think that everything here is the work of a devil. The rabbinical tradition, probably in order to outdo the Persians, went further and claimed that God created seven Heavens and seven Earths. As much as they confused the concepts of the heaven and of the universe, they wanted desperately to appear right, and to support their claim to such an exclusive knowledge, they even provided us with names for all these Earths: Eretz, Admah (i.e. Red!), Arka, Harabah, Yabrashah, Tevel, and Heled. The last one, according to this tradition, is our very own planet. Turning our attention to the word Arka, so close to the fabled 'ship' Argo, we ought to realize that at least three of these names refer to the ancient name for shining white metal: "Arg." We find this reference, albeit in an altered fashion, also in the names Harabah, and even Eretz.

A whole additional store of enigmatic names can be culled from Jewish lore, courtesy of an old, and very insightful book titled *Oedipus Judaicus*. For example: Bith Joshimoth means a House of Heavens, Bith Hogla is the House of the Circle, Bith Aneth is the House of Affliction, and Bith Shittah is the House of Declination, which, as we shall see shortly, has its counterpart in Celtic mythology, where it refers to one of the Revolving Castles.

There are other enigmatic Hebrew expressions available: Adami Nekeb means "the red ones of the hollow, or concave space." Remon-methoar is translated as "the high place, or city encompassed by a circle." Remon-methoar-neah could then best be translated as a "circular rotating city," since the expression "neah" implies motion. Furthermore, the well known Jewish name Beersheba can be translated either as "the well of oath," or as "the luster of the seven."

The Egyptians have also preserved some rather intriguing names in their mythologized lore. For example, their crocodile god Sobek is also known as the Lord of Bakhu, the mythological 'mountain' of the horizon, where he had a temple made of carnelian, a cherry-red quartz, which should remind us of the often mentioned 'red' satellite. Perhaps even more intriguing are references to Osiris, who was purified on the day of his birth in the great double nest of Hensu, likely the "up and down" tiered space habitat. Although Hensu is the Egyptian name for the city the Greeks called Heracleopolis, nests, especially in mythological contexts, are usually associated with flying. It is in this association, two other 'nests' are known: the nest of Millions of Years, and the Great Green Sea nest, where Ra was supposed to live. (Probably the same 'Sea' others called the River Ocean.) In the same 'sea' most likely was located the Island of Maat. The lore also speaks of three mysterious 'gates.' The Gate of the South, the North Gate of the Domain, also known as the Tomb of the God, and the Gate of the Pillars, also known as the North Gate of the Tuat. Again, we have seven names here, of which the last one most likely represents the space elevator. But since the designations are so nebulous, as is so much Egyptian mythology, we merely present these, and the Jewish names, as tantalizing possibilities.

On the other hand, the information I was able to reconstruct, concerning the Celtic Revolving Castles, is quite definite and specific. I have to thank Mr. Robert Graves here, for presenting this lore to us in his *White Goddess*, and to the Welsh people, to all the bards and the knowledge passed on by Druids, who preserved this memory for us.

Presently, it is the round burial mounds, found in England and elsewhere, that are referred to as the Revolving Castles, but, of course, it was not always so. While a lot was forgotten, some amazing details were remembered. In his poem, the *Reproof of the Bards*, the famous 6th century bard named Taliesin ("Fine Value") taunts the leader of the bards, Heinin, for not knowing: "The name of the firmament, the name of the elements ..." To prove his value Taliesin provided these names for us in the composition titled *The Spoils of the Bottomless Place*. (Preiddeu Annwm.) Here he names altogether eight such places though: Caer Sidi, Caer Pedryvan, Caer

Vadiwid, Caer Rigor, Caer Wydr, Caer Colur, Caer Vandwy, and Caer Ochren.

According to the Welsh tradition, Caer Sidi alone means a Revolving Castle. Caer Pedryvan means a Four-Cornered, or Four-Times Revolving Castle, Caer Vediwid the Castle of the Perfect Ones, Caer Rigor the Royal Castle. While Caer Wydr remains without an epithet, Caer Colur means the Gloomy Castle, Caer Vandwy the Castle on High, and Caer Ochren the Castle of the Shelving Side, which Robert Graves explains as being entered from the side of a slope. (Re: the rabbinical House of Declination.)

Because of his desire to spotlight his knowledge of the ancient lore, Taliesin, in order to keep the number 'correct,' did not count the Caer Arienrhod among the firmaments, because the name was assumed to mean Silver Wheel, which he most likely mistook for a generic name for all such structures.

However, when we consult the Indo-European dictionary, that goes back to the roots of Western languages, as they were being used 6,000 to 9,000 years ago, we discover different meanings for the expressions given above, which of course are much older than the Gaelic practiced by the Welsh bards of the 6th century.

In the Indo-European language, where the preceding 'original' meaning was already transformed, the word "Caer," if translated literally, could be rendered either as "round-shaped shining metal," or as "to fit together shining metal." i.e. a constructed firmament. The word "Sidi" or "Side," most likely meant a "bound" or "filial seat." Freely translated the two words then can be rendered as a "shining (metallic) orbiter." In other words, in this linguistic context, Caer Sidi is the generic name for the near-Earth space habitats that we call here Revolving Castles, or Firmaments.

On the other hand, when we analyze the etymological origin of the name Arienrhod, we realize that its primary meaning was "Red Rule," but it could, at the same time, be also perceived as a "Conch-shaped" or "Spiral Shaped." And having named these two qualities, we should recall the tiered "Red House" of the Beloved Gods from Hopi mythology, the Mayan "Red Mountain," the Hephaestus' mansion in the sky, made of 'copper,' etc. (We should also point out that the tiers of such a spiral would appear like 'steps.' (Re: ziggurats [and stepped pyramids], the 'abodes of gods' on Earth.)

The problem with the Indo-European language, from our perspective, was the fact it was agglutinative, and the resultant information was often multi-leveled – which can also tell us something about the mental state of those people. However, as the ability to use this language declined, meanings started to overlap, and diverge as well. For example, the key component for the word Caer, the "ker" sound found in the dictionary, has about five different meanings, including horn, head, grow, heat, etc. A related word "kerd" means "heart," and the Mayans speak quite often in their myths of the "Heart of the Sky." In the Orphic texts we come across an argument attempting to explain how a class of angels known as "thrones," (i.e. seats), actually should be called "revolving." Obviously, the agglutinative language, after its demise, caused a lot of confusion for the succeeding generations of translators.

And it is causing a problem for us as well, because the "ker" sound simply does not fulfill our need here, and we have to consider another root word with a close audio affinity to "ker," which happens to be the word "gher." It was assigned all the following meanings: a gut or entrails, (it 'spirals' too), to grasp or enclose, with derivatives meaning of "enclosure," as well as "to shine," or "to glow." Most of these meaning do fit the context.

When we turn our attention to the root-meaning of the other named firmaments, the most dubious claim to existence is presented to us by the Caer Rigor, because the majority of the root-words for Rigor suggest that it was destroyed. But the combination of the root-words "re + ghēu + orbh," i.e. "gape-back-asunder" strongly suggests it was in fact the Space Elevator, severed from its Earth-side base, gaping down from the sky. We shall see further confirmation of that in the next chapter. Another root word, "rei," as in "rei + ghēu + orgh," refers to a band of various colours, as in the English word "rainbow." But the 'lid' of Sampo, the Finnish version of the Space Elevator, was also said to have been of many colours. All in all, we can safely discount Caer Rigor as being one of the Revolving Castles. It was the Space Elevator in its damaged state, which also means that the information itself must have been hoary with age.

Of the following three Revolving Castles we can equally safely say that, judging by their root words, they were already destroyed and, as we shall see, destroyed in remote prehistory. The three include Caer Vandwy (Satellite is Gone), Wydr (Split Apart), and Colur (Destroyed Overhead). The one with a question mark is Caer

Vedivid. This name can mean either of two things: "The Creative Force is Gone," or "Water-Merged." The second meaning can, in turn, have two explanations: 1. Most likely it crashed on the planet's surface, and 'merged' with the waters of an ocean, or 2. It started falling, but its descent was (temporarily) halted on the level of the River Ocean, i.e. the upper reaches of the atmosphere. Descriptions of both events have survived.

The one remaining name, Caer Ochren, simply means "True Satellite." (The "O" means "to announce," or "to hold as true" and "Ghren" means "to grind.") The last name on our list, Caer Pedryvan, is not a Revolving Castle, but, again, the (same) space elevator, this time in proper working condition. The "ped" part means either a foot, (re: the Mayans), or a container. "Dreibh" means to drive, or to push, and "wen" to desire, to strive for. Freely interpreted it can be rendered as "desirable transportation through a column."

What this Welsh account tells us is that, indeed, there were six space habitats and a Space Elevator. It also suggests, as do some other sources that, possibly, the Space Elevator might have been four sided instead of round, but that too could be a matter of misinterpretation of the information available. Confusing perhaps the three supports and the main structure for the four sides, or they might have confused it with the four cardinal points of the world. To underline the uncertainty of such secondary assumptions, I should point out the Celtic people also used to maintain lore of eight Islands of Earthly Paradise. The silver one, named Findargad, was believed to be supported by four pillars, which would point to a yet another concept of the space elevator and to the origin of the Egyptian idea of four pillars supporting heaven. But there was also the Island of Ildathach, the Many-coloured, which would again point to the space elevator. After all, there was only one such fabulous structure upon the planet, and fuzzy facts would have engendered many variations on the theme.

While we are still dealing with the topic of Revolving Castles, I would like to bring your attention to the word "Sidi" or "Side," as it traveled around the world. The 2,000 year old book on 'primitive' astronomy, known in India as *Surya Siddhanta*, which itself reflects the name, tells us that "below the Moon, and above the clouds," which more or less defines a Lagrange point, "revolve the Siddhas," or "perfect men," and the "Vidyaharas," possessors of knowledge.

(But to be perfect requires knowledge as well.) From the Japanese tradition we also know that everything perfect was considered divine.

In ancient Greek "Sideris" was the word for a star, and "sideros" for iron. In Lithuanian the word "Suidu" means "to shine." Also, the beings between angels and men were called by the ancient Hebrews "Sadaim," and by the Greeks Daimonas. Yet among the ancient philosophers these 'demons' were deemed to be an aerial race. The Finnish Kalevala also mentions "a race of Sindri," living way UP north.

THE WAR OF THE GODS

"And He changes the times and the seasons." (Daniel 2:21)

The indication from the previous chapter is that only one Revolving Castle, Caer Ochren, is possibly still in orbit around the Earth, invisible to our eyes, and to our instruments. However, such possibility is quite remote, considering all the aspects of the evolutionary role these structures, and their inhabitants, played in the life of this planet. While I am fairly certain the descendants of those people have extensive basis on this planet, as well as elsewhere in the solar system, and that most of the "UFOs" originate from such bases, I am equally certain that all the revolving castles are gone, by now.

They were eliminated from the scene in two stages, the first of which was represented by an all-out 'nuclear' war. While it is usually referred to in myths as "The War of the Gods," indications suggest one party in that conflict was much more concerned with the evolutionary progress of the planet-bound humans than the other. But such suggestions too can be misleading. All that the myths have bequeathed to us is a few shocking and revealing pointers that enable us to reconstruct only the most basic facts about this exceptionally destructive event, and no more.

Yet, this horrendous conflict, along with its even more horrendous consequences, must have been anticipated by at least one of the parties in the conflict, because a Persian myth tells us that Ahura Mazda, the supreme God, ordered Yima Xšaëta, the King of this world, to construct a Vara, or subterranean enclosure. We are told this Vara was a cube, each side of which was as long as a race track, or about 500 meters, or 1,500 feet. "There he brought the seeds of sheep and oxen, of men, of dogs, of birds ... the seeds of men and women (were) of the greatest, best and finest kind on this Earth ..." and also "the seeds of every kind of cattle ... of every kind of tree ... to be kept inexhaustible there ..."

In other words, Yima had created a large genetic bank. But it was more than that, because he also situated there a 'town' of 1,900 men and women, complete with streets, courtyards and balconies, complemented by a flowing river with "evergreen banks, that bear never-failing food," all illuminated by "a window self-shining within."

Reading the text I was left with the impression that either there were in fact three such Varas, or that the one was divided into three sections, each with its own human crew, the largest of which, counting 1,000 men and women, was to look after the genetic bank itself, 600 to look after the hydroponic gardens and living animals, and 300 to keep records, and to provide various other services. In this enclosure a couple was allowed to have one boy-child and one girl-child every 40 years, to maintain a steady population.

This story has its counterpart in North America, retained for us by Wintun Indians of California. A being known as Olebis (He-Who-Sits-Above) informed the Wintuns: "There was a world before this one in which we are now. That world lasted a long, long time ..." But it was destroyed by fire. "Olebis looked down into the burning world. He could see nothing but waves of flame ... great rolls and piles of smoke were rising ..." Yet a day before the fire started, we are told, Olebis and two old women built a huge "sweat-house," and "the old women gave him all kinds of beautiful plants now, and flowers to form a great bank around the bottom of the sweathouse ..." Then Olebis said: "I want this house to grow, to be wide and high, to be large enough for all who will ever come in."

The War itself most likely started with a sudden attack against one of the two trios of space habitats, and I would suspect the trio above the Antarctic was attacked, simply because there was presumably less traffic there, and possibly less surveillance against an attack from the Earth, than there would have been in the north, where the space elevator was located. The other reason will become more apparent as we move along the story line.

We can piece the event together from the bits and pieces scattered throughout the Indian sources, including several translations of the epic *Mahabharata*: The "all-traversing automated flying vehicle ... reached the favourite divine place of Indra ... (with its) thousands of flying vehicles ... When the three cities in the sky met, god Mahadeva pierced them with his terrible ray of three attack bands," and "a(n) incandescent smoke, like 10,000 Suns, rose in all its splendour," proclaims the sentence made famous by J.R. Oppenheimer, the 'father' of the nuclear bomb.

The Gnostic sources comment on the same subject: "As to the fight by the Persae Tree, hard by in Annu, it concerneth the Children of Impotent Revolt, when justice was wrought on them for what they have done. As to (the words) 'that Night of the Battle,' they concern the inroad into the eastern part of heaven, whereupon there arose a battle in heaven and in all the Earth." Obviously, the above spatial directions are confusing. While the Persae Tree is the space elevator, and Annu here refers to the Firmaments, how do you define 'east' in space? My assumption is that the reference is to the three space habitats located in L1 would be nominally in the 'west,' that is above the North Pole, and the L2 satellites would follow in the "east," from their position above the South Pole.

The theme of 'retribution' echoes even in the Bible: "They came from a far country, from the end of heaven, even the Lord, and the weapons of his indignation, to destroy the whole land." (Isiah 13:5) What the nature of this "indignation," of this "impotent revolt" was, shall be touched on in the next chapter.

In the Gnostic texts we find the following: "The heaven and his Earth were destroyed by the troublemaker that was below them all," that is on the surface of the planet. "And the six heavens shook violently ..." Actually, three were vaporized, and two went out of their orbit, as another source informs us. Understandably, the survivors were beside themselves with sorrow and rage, which they swiftly proceeded to unleash.

The *Mahabharata* retells the story this way: "The valiant Adwattan, remaining steadfast in his Vimana," (a space-going ship), "landed upon the water" (of the River Ocean, i.e. the upper atmosphere), "and from there unleashed the Agneya weapon ... taking aim against his foes, (he) ... let loose the blazing missile of smokeless fire with tremendous force. Dense arrows of flame, like a great shower, issued forth upon the creation, encompassing the enemy ... all points of the compass were lost in darkness. Fierce winds began to blow. Clouds roared upward, showering dust and gravel ... the Earth shook, scorched by the terrible violent heat of this weapon. Elephants burst into flame and ran to and fro in frenzy ... the waters boiled and the creatures residing therein also died. From all points of the compass the arrows of flame rained continuously and fiercely. The missile of Adwattan burst with the power of thunder, and the hostile warriors collapsed like trees burnt

in a raging fire. Thousands of war vehicles fell down on all sides."
(I.e. thousands of aerial vehicles.)

Furthermore, the account also tells us the corpses were burnt to ashes, survivors' hair and nails fell out. Pottery broke without cause. Foodstuffs were poisoned. To escape, the warriors threw themselves in a stream to wash themselves and their equipment. All clear signs of a nuclear war. Unless of course, high energy weapons akin to lasers, lead to the same consequences.

We can also see the reflection of the same event in Hesiod's *Theogony*, in the part describing the consequences of the Titans' revolt: "Then Zeus no longer checked his rage ... He came from heaven and from Olympus, lightning as he came, continuously; from his mighty hand the bolts kept flying ... while the holy flame rolled thickly around. The fertile Earth being burnt, roared out, the voiceless forest cried and crackled with fire; the whole Earth boiled, and the ocean streams ... flame unspeakable rose to the upper air ... the awful heat reached Chaos ... the howling winds brought on dust storm and earthquake."

But it is not just these three sources. The images of repeating 'thunderbolts' and of burning Earth are present in most mythological systems, all over the world. Different parts of *Mahabharta* call it either an unknown weapon, or Vajra weapon. In Sanskrit though, vajra means a diamond, which would suggest a type of laser. The Norse god Thor used a 'hammer' for the same purpose, the Phoenician god Baal 'clubs'. The Mesopotamian war god Ninurta also used 'clubs' or 'cudgels,' known either as "world crusher," or "world grinder." In either case, the Mesopotamian myth of Anzu makes it fairly plain that Ninurta represented the interests of the planet-based powers, who challenged those of the heavens. The outcome, by now, sounds familiar: "Clouds of death rained down, an arrow flashed lightning." Surprisingly though, we also learn in this story the sky 'gods' were able to override the programs of the weapons fired at them by the Earth-based 'gods'. Ninurta "aimed the shaft at him ... but it did not go near Anzu: the shaft turned back." Anzu, it turns out, "was holding the God's Tablet of Destinies in his hand, and they influenced the string of the bow ... Weapons stopped and did not capture Anzu amid the mountains." Yet, later "a heatwave blazed, confusion ..." In the end "Ninurta slew the mountains ..." The three 'mountains,' or space habitats, we spoke of earlier. This, most likely, is a typical mythological reversal of events, but we cannot really be sure here either.

Among the Chinese sources the story of "Shen-I, the Cosmic Archer," stands out among the more instructive ones. In one of his 'adventures', the hero encounters nine fierce phoenixes, who "issued forth tongues of fire from their beaks, forming nine false Suns whose rays threatened to destroy the land ... but Shen-I wisely ignored the phantom Suns. Instead he saved his arrows for the throats of the nine birds themselves. As the throat of each bird was laid open by an arrow, the false Sun above it dissolved into a mushroom cloud." Afterwards, his troops found nothing but nine red stones. We can only presume these stones were rubies, used to generate a laser beam, and that the weapon was powered by nuclear energy.

The *Bhagavata Purana* bears this record: "The flames of the Brahmastra-charged missiles mingled with each other, and surrounded by fiery arrows they covered the Earth, Heaven and space between, and increased the conflagration like the fire and the Sun at the end of the world ..." Even the Australian aboriginals remember that "two brothers came down to the Earth world to hunt," from the sky world above, where they lived, "but their firesticks started fires all over." The Celtic tribe of Boiims also claimed that "the round land was burning." The Voguls of Siberia remember somehow that "God sent a sea of fire upon the Earth ... For seven winters and summers the fire raged."

In other words, this insane war possibly lasted seven years, unless a 'sacred' number was used here, to mask the absence of a factual knowledge and, as the mythologized rendition of it, in form of the fight of Seth and Horus suggests, the war raged on land, on the seas, as well as underneath them, before it culminated in the conflagration that afflicted the entire planet. A conflagration that would have been made worse yet by the impact of the burning debris of the space habitats, of the destroyed weapons, of space and aerial vehicles.

"The Earth trembled, and the heavens dropped ... the mountains melted," is stated in Judges 5:4,5. The Ovaherero tribesmen of East Africa say: "the Greats of the sky (Eyuru) let the sky fall on the Earth; almost all the people were killed." The Cashinaua tribesmen of Western Brazil intone: "The lightning flashed and the thunders roared ... then the heaven burst and the fragments fell down and killed everybody and everything." The Nordic Edda asks: "What happens when the whole world has burned up, the gods are dead, and all of mankind is gone?"

Of course, with this level of destruction, the space elevator did not remain unaffected. A Chinese account of its partial destruction, quoted here, probably comes as close as can be expected to the actual event: "... the pillar cracked and fell in pieces, broken, cutting great rifts in the Earth." Then "the sky collapsed and crashed down ..." (I.e. the Revolving Castles crashed.) Another story tells of a divine couple who "lived on Fusang, an enormous tree which towered a thousand zhang (about three kilometers) into the air from the depth of those boiling waters." (Of a sea called Yanggu.) The Greeks present the destruction of the space elevator as the "separation of Uranus (Heaven) and Gaia (Earth)." It is a forgone conclusion the sky people had won, but they were sensible enough to save a great number of Earth-people, including their arch-enemies, the giants, inside the remaining space habitats, and the safe sections of the space elevator.

In the end the devastation was utterly thorough and the Earth floated in space lifeless and sterile. Although mythologized records of the subject differ, it remained lifeless for a long time.

An Amerindian story tells of three brothers who stayed in the upper world for fifty years, while no one had lived in the villages down on Earth. Most likely they also died in the upper world. An Innuit Creation Myth states this: "The land they came from he (the Raven) called Heaven ... Here everything was deserted and sterile, and he again planted ... this new land Earth."

The Persian Creation Myth offers us one possible time segment, during which the Earth was 'dormant'. It states that Aharman made "a treaty of nine thousand winters ... with Ahura Mazda ... and when the nine thousand years have become completed, Aharman is quite impotent ..." Aharman, in this instance, is presumably the dangerous level of radioactivity. An indirect remark in the *Mahabharata* more or less agrees with that figure: "... it is harder by ten million times to call back the weapon once released, and at the slightest error ... (the) Earth (would) become desert with no life for seven thousand years." Unfortunately, both of these figures again are tied to the "sacred numbers," and thus not quite reliable.

The research of contemporary sciences concludes that no finds exist for carbon dating for the period between 25,000 and 21,000 years ago. In other words, there was no life on this planet for at least 4,000 to 5,000 years to leave any traces for us to find. Furthermore, vastly excessive levels of dust, discovered in ice cores, were dated to between 25,000 and 24,000 years ago. This could hardly be there as the result of the subsequent maximum glaciation and desertification of the planet, and of extreme storms, (that came somewhat later), as is usually claimed.

It is said quite plainly in the *Ramayana* that deserts were made by the terrible weapons of the gods. As if to confirm the fact, layers of fused glass, that we know are caused by nuclear, or other high energy weapons, have been found in these deserts: Mohave Desert, Gobi Desert, in southern Iraq, in Palestine. Finds of fused and scorched forts, or cities, were reported from India, Brazil, West Pacific, Norway, and elsewhere. But, admittedly, some of the later findings need more support for such a claim.

Intriguingly, it was also realized recently that Earth is surrounded by a thin layer of anti-matter, of all things, left behind by the destroyed Revolving Castles. But the scientists, who know naught of such events, are suitably puzzled.

A NEW CREATION

In general, I would date the period of The War of the Gods to between 25,000 to 26,000 years ago – right into the period of the Age of Kali. The confirmation, again, is to be found in the *Mahabharata*, which states of the time of this conflict: "The era of Kali has arrived, when the laws of the previous age cannot apply."

While the culmination of such era is upon us again, we shall not be repeating the same fatal mistake. The mistake specific to our age is releasing vast quantities of chemicals into the air, water, and soil, as well as vast quantities of carbon dioxide, which is going to cause a great cataclysm and population loss, but certainly not on the level attributed to The War.

Scientists, such as Carl Sagan and Isaac Asimov, used to worry mankind will commit a suicide by unleashing a nuclear war, mainly due to our immaturity as a species. It was even calculated that only a tiny fraction of sentient species, who ever existed in the universe, would survive this particular stage of their evolution. To prevent such an outcome, Isaac Asimov was proposing, in his science fiction novels, that the immature mankind be perpetually guided and protected by sentient robots.

Such a proposal though, in spite of its humaneness and generosity, sprung from the gentleman's materialistic conviction that the entire universe, and all the life in it, including all sentient biological life, is a mere accident. As we have surmised by now, and as I started to demonstrate in the book *Nostradamus Decoded*, that certainly is not the case. It is in fact the Supreme Being that is, and always was, the guide and protector of all life, and will remain so, as far as this planet is concerned, until our species matures – which stage is fairly close.

If mankind was allowed to wage such an extremely destructive war, it was to do so under the conditions that would guarantee the renewal of life on this planet, and only in order to accomplish several objectives: 1. Imprint the horrors of it on our racial memory, in order to prevent us from starting such a war ever again. 2. Remove undesirable, anti-life species. 3. Return the sky people back to Earth.

Of course, returning the sky people to Earth was a long, drawn out process that took about 20,000 years to accomplish. For starters, the level of radioactivity had to decrease to an acceptable level, before the terraformation of the planet could begin. Although the planet was probably terraformed by the sky people at least twice before, roughly 100,000 and 250,000 years ago, never before the planet had to be brought back from the dead, and while in the grip of a nuclear winter, with much of its land mass covered in ice.

Yet, in spite of the utter destruction the sky people had inflicted upon the planet, they did not pack up and abandon this world to go somewhere else. The reasons for that are quite simple: There are not that many life bearing planets out there, and those that are there most likely harbour an intelligent life of their own. Also, the very nature of this universe virtually guarantees the same predicament of climatic swings, and of other natural cataclysms, as Earth does.

This, however, was not merely a matter of theory for the sky people. They had first hand experience in such matters, which is attested to by references, in several myths, to spaceships and space travel. While several of them mention "travels from one planet to another," these could mean travels within the solar system, or simply the short hops from one space habitat to another, since these were also referred to as 'planets,' or even as 'stars,' by the Earth people.

Instead, we shall concentrate here on the trips to the distant stars and galaxies. In this context, we can perhaps take the following exclamation, attributed to the Egyptian god Horus, as a true indication of such travel: "I shall annihilate the devils; those who travel through the sky … and also those who reach the stars." A Chinese story mentions the hero Shen-I, who was "taken by a streak of light across the stars," which, I think, refers to a spaceship traveling near the speed of light.

A somewhat more tangible evidence of intragalactic space travel comes to us from the Ongwe Creation Myth of the Iroquois people. The Ongwe were beings who lived on the "other side of Heaven." One of them 'died' and was placed in "the coffin on a high place." His young daughter cried and cried, until she was "lifted up," several times, to see her 'dead' father. She was also given a 'ring' by this 'dead' man, which she then used to communicate with him, for years, in his 'coffin'. Then, at his suggestion, she traveled across the Milky Way, i.e. the galaxy our

71

solar system is located in, to marry a young chieftain. And she made this journey three times, we are told.

While most of the story is inevitably a confused gobbledygook, the pointers it provides are crucial. Namely that intergalactic voyages, initiated by the sky people, did take place, and that they took years, even at the speed approaching that of light. As if to confirm this fact, stories of time dilation exist in China, Japan, Middle East and Bohemia, and were recorded even in the Egyptian Book of the Dead: "… (he) strides over distances of millions of years in one little moment." This certainly represents an exaggerated opinion, as far as our understanding of such matters is concerned, but retains the gist.

The size of these 'coffins' was quite awesome. One of the Pyramid Texts mentions that Horus' ship as 770 cubic feet long, which is about 2,000 feet, or just under 700 meters. The Mayans claim that the Golden Ship their ancestors traveled in to the stars was about six kilometers, or almost four miles, in circumference! The Finnish Kalevala mentions a flying ship with a hundred men crew and 1,000 passengers. The Nordic myths say of Odin's Golden ship, or of Freyer's "cloud ship" Skithblathair, that "she is so great all the Aesir … may find room on board her." Considering that the Nordic Aesir are the same as the Hindu aerial race of Asir, and that they are both identical to the Sidim of the Revolving Castles, this claim, no doubt, represents a gross overstatement of the fact, but not of the reality it was once derived from.

Returning to our main theme, the consequences of The War that destroyed the then extant life on Earth, we have to accept as a fact that only a relatively small group of Earth-born survivors, counting perhaps tens of thousands, in total, could have been rescued from the planet, and then 'stranded,' for millennia, in space. It was probably at that stage the sky people realized that they, after circa 300,000 years, are no longer genetically compatible with the surface people. To ensure their future survival on the planet, where they knew, by then, most of their descendants will eventually end up on, the sky people started to take steps to rectify the problem, and after enough time had elapsed, the stage of forced return to the planet of their origin was upon them.

The (Chinese) annals of the *Bamboo Books* carry this statement: "… the five planets went out of their courses. In the night, stars fell like rain. The Earth shook." The dating provided for this event, the

10th year of the Emperor Kwei rule, would place it to about 2,000 BCE. The event, as described, is essentially true, as we shall see later, but a much older information was incorporated in it, namely about the five 'planets'. In my opinion, this refers to the state of affairs, almost 26,000 years ago, when three such 'planets,' i.e. space habitats, were destroyed and additional two most likely were knocked out of their orbit during The War, but eventually were able to regain it, for several more millennia. In either case, as we shall see, the Chinese knew five immense structures fell, and most likely saw one of them falling themselves.

The Athabascan Indians, for example, have a tradition that tells of gods repairing the skies, as they were about to fall. From India comes the myth of the universal king, the Wheel-King, named Strongtyre, "who possessed the seven precious things." One day the wheel, whose steadfastness was the measure of his fitness to rule, began to slip. The King then called his eldest son and spoke to him: "Behold, dear boy, my Celestial Wheel has sunk a little, has slipped down from its place ..." The story mentions the ultimate demise of the Wheel, most likely thousands of years later, and we shall return to it at an appropriate time.

Considering that at least two out of the three remaining space habitats were doomed to eventually crash onto the planet, their original inhabitants, with the prospects of being forced back to Earth, had four crucial problems on their hand: 1. Keep the habitats aloft as long as possible. 2. Repair the damaged Space Elevator. 3. When the radiation falls to an acceptable limit, start terraforming the planet. 4. Deal with the genetic incompatibility with the Earth-born people. (After 300,000 years the two became separate species, incapable of mutual procreation.)

While we do not have any stories about the post-nuclear terraformation of this planet, we do know that the (Canadian) Arctic was strangely warm about 20,000 years ago, and that trees grew there, while Amazonia, virtually on the equator, was a cold savannah. The most likely reason for that was the different position of the polar ice masses, but it is also quite conceivable that this part of the Arctic was artificially heated, perhaps by ultrasound from the space elevator, as are some modern hothouses.

As the Earth was gradually warming-up, and the areas of tree and vegetable growth started to expand, it was possible to reintroduce other life forms to the planet from the Vara(s) and, mainly, from the space habitats. To illustrate; one story of the Tibetan creation myths mentions the animals brought from the other-worldly "Naga realm," and a question is asked whether they have freshly given birth, obviously expressing safety concerns about the effects of radiation. The Amerindians are more direct. The Tsimshians, of the Northwest Coast say: "A giant scattered salmon and trout into Skeen River." The Plains Comanches maintain that "long ago two persons owned all the buffalo ... they kept them penned up in the mountains ..." The Chinese also believe the ox was sent from heaven, and the Celts that the pigs came from Annwn, the Bottomless Land. The Dogons of Africa believe the "heavenly smith" brought grain and agriculture to Earth.

Within a few thousand years most of the available land surface area was covered with trees, shrubs and other vegetation, or at least with grass, as well as with animals of all sorts, appropriate to their place or origin, including the mega-fauna in the Americas. Human beings, native to the Earth, also started to resettle the planet, from their long sojourn inside one of the space habitats, or inside the remnants of the space elevator. As several stories say, or suggest, the "surface people" were accepted there as 'refugees'. Since the living conditions would have first become suitable for life in Southeast Asia, most people would have probably decided to live there. The European population numbered only some 16,000 people, during that time, according to the current estimates.

But the sky people, knowing that their descendents too will have to eventually abandon the space habitats, and return to the planet of their origin, had also realized that to do so in their present state would sentence their race to extinction. If they did not become genetically compatible with the surface people, their small, isolated groups, conditioned for eons to maintain steady population, perhaps even genetically programmed to do so, would soon perish on the planet, which they had no intentions of doing.

On the outset we should mention the myth native to both India and Iran. It is about the Premordial Man, named Gayomart, whose body was believed to have been made out of *seven* kinds of 'metal'. From his seed, purified by the *rotation of heaven*, was born the first human couple. If it needs stressing, in this context, the words seven, metal, and the rotating (or revolving) 'heaven' provide the needed clues. The only problem with this story is that, as usual, only a part of it is correct.

The Zuni Indians, on the other hand, believed humanity was born in the deepest of *four* chtonian "cavern-wombs." At least their count reflected correctly the real post-War number of usable space habitats: three Revolving Castles and the Space Elevator. As far as the claim for such "cavern wombs" is concerned, it can be confirmed by the quotes from the Mesopotamian myth of Atrahasis.

"... Let the womb-goddess create offsprings ... they called up the goddess ... the womb-goddess (to be the) creator of mankind! ... (She) made her voice heard and spoke to the great gods, 'it is not proper for me to make him, the work is Enki's; he makes everything pure! If he gives me the clay, then I will do it.' "

Enki agreed and started a purification process. Then one god was 'slaughtered' and the womb-goddess was to "mix clay with his flesh and his blood. *Then a god and a man will be mixed together* in clay. Let us hear the heaertbeat forever after, let a ghost (of intelligence) come into existence from the god's flesh ..."

Obviously, the Mesopotamians knew nothing of genetic manipulation, and would not have been able to imagine not only the DNA, but that such could be 'mixed' in anything but clay. The clay they were certainly familiar with, since their homes, furniture, utensils, writing tablets, etc., were all made of clay. As far as the "birth-goddess" is concerned, she had a host of other names as well, and these translate as: mountain lady, birth-lady, mistress of the gods, midwife of the gods, even a "smelter of the gods," which is probably the one most appropriate, since two different kinds of DNA were 'smelted' together in 'her' to produce one. Obviously this 'goddess' was nothing else but the equipment of massive genetic labs, along with the required artificial wombs. Altogether fourteen 'womb-goddesses' were assembled, as the story plainly states, and at the end of the term these were opened with a staff, or lever.

We can hear an echo of this theme in a Welsh poem *Diversified Song*: "She was self-bearing, the mixed burden of man-woman." These words allegedly apply to Eve, the "mother of mankind." But the 'womb-goddess' was also known as the "mother of all children," and thus of all mankind. The Gnostics also claim that Eve bore her first offspring without a husband, and served as her own midwife, just like the 'womb-goddess.' And just like the 'womb goddess' she also was "the process of becoming." The text further reads: "The souls that were going to enter the modeled forms of the authorities were manifested to Sabaoth ... And regarding these the holy voice said, "Multiply and improve!' " But Sabaoth is merely a name of one of the seven 'offsprings' of Yaldabaoth, the Space Elevator, as was explained earlier. This particular space habitat probably made science its business. There are even hints The War started because of the activities there, and that it was also the one that survived (the longest).

The Gnostic texts are also much plainer than others on the need for such action: "He (the chief archon) sent his angels (i.e. messengers), to the daughters of men, that they might take some of them for themselves and raise offsprings for their enjoyment, and at first they did not succeed. When they had no success, they gathered together again and they made a plan together. They created a counterfeit spirit, who resembles the Spirit who has descended ..." That is, they adjusted their DNA to fit better the human one, and vice versa. Another Gnostic text reads: "His (man's) modeling took place by parts, one at a time." And the sky people's "leader fashioned the brain and the nervous system." But what could the Gnostics have known about such things? Aristotle, a virtual contemporary of such ideas, who also 'borrowed' some, thought the brain was a cooling organ of the human body, analogous to the radiator in a car. Only tradition maintained these 'stories'.

The above idea also finds support in the Mayan *Popol Vuh*: "... our first mother-fathers. They were simply made and modeled, it is said: they had no mother and no father." However fantastic it may sound, only such genetic modeling can explain the mystery of the "genetic Eve," i.e. the mitochondria, whose origin can be traced to one woman who lived about 250,000 years ago, but is shared today by all humans, and of the "genetic Adam," dated to circa 100,000 years ago.

THE FALL

By about 16,000 years ago the world was again viable, and where conditions permitted, fairly densely populated. Even the 'myths' can tell us how fast the repopulation can progress. For example, the first races, most likely of the mixed-breed origin, to settle in Ireland, grew from 48 people to 5,000, in only 300 years. Theoretically, in another 300 years there would have been over 500,000 of them, and another 300 years later still fifty million. The Mesopotamian epic of Atrahasis concurs: "600 years, less than 600 years passed, and the country became too wide, the people too numerous." That is what would have happened if a few thousand people came, and the same incremental ratio was maintained, as in Ireland, for less than 1,200 years.

The speed of expansion for any population is regulated by two simple factors: 1. Availability of productive land and resources. 2. Competition for the same. Since the land was empty of human population and, again, 'virginal', and the new population was under direct command to reproduce as much as possible, the population could have easily reached tens, even hundreds of millions, within the said 1,200 years, in the repopulated regions of Southeast Asia.

But then, just about 15,500 years ago, a cataclysmic convulsion ended the Ice Age proper. While it is possible this event too was triggered by the sky people, and the Mesopotamian myths quite often mention a "flood weapon," along with a similar accusation found in the Bible, these are open to a debate, because most of the global myths do not view the flood as a punishment. That, in itself, does not mean much, because only 2,900 people had survived the cataclysm, a Persian source tells us, and they were in no shape to remember anything. Among the three races that perished were four million giants. And the giants, according to the Greek myths, were the principal adversaries of the sky gods in the past.

How old such animosities might have been is also debatable, because remains of giant human beings have been found that predated the Homo sapiens. They could also be the Earthbound troublemakers, some sources referred to, whose evolutionary demise came due as well. The last we hear of them is in North

America, where the Cherokees drove what was left of them from their walled towns, and sent them fleeing westward, where they eventually perished, about 1,000 years ago.

In the wake of this latest cataclysm, the sky people, now also hybridized, again had to repair the devastated planet, reintroduce both culture and agriculture to the decimated remnants of humanity, who either sunk below the animal level, to the point of forgetting the spoken language, or resorted to cannibalism. Several Asian 'myths' have passed that information to us, in fairly gruesome details. At the same time though, a university was (re)established somewhere in Asia, according to a Tamil 'myth'.

The Mesopotamian 'myths' devoted to that period speak of the Igigi, the "younger generation of gods," that is the 're-engineered' sky people, now sexually compatible with the "surface people," who carried out the workload of terraforming the planet, by digging out new channels for rivers, by draining huge marshes, by planting forests. We are told that after 3,600 years of heavy labour, the Igigi 'rebelled', and took their complaints to heaven. But most likely they simply had to interrupt their work, because of the worsening climatic conditions. The climate started to fluctuate and then, somewhere around 11,700 years ago, the global temperature shot up by at least 7 degrees Celsius, but more likely by about 10 degrees, within a couple of decades, and stayed that way for some 1,500 years. This made the conditions over most of the habitable parts of the planet so intolerable the survivors were forced to migrate to the northern polar region. The early stages of this situation too are captured in the Hittite 'myth' of *The Vanished God*, which plainly states it got so hot all mammals lost the ability to procreate. This could have happened only if the global temperature exceeded 24 degrees Celsius.

Then, about 10,200 years ago, an asteroid appeared in the sky, heading for Earth. It was a massive one, about ten kilometers in diameter. It impacted the Earth, in a fragmented form, along the eastern seaboard of North America, with most of the fragments impacting the submerged continental shelf. But, as another Hittite 'myth', *The Song of Ullikummi*, attempts to explain, the fragmentation did not take place spontaneously. The story recorded on clay tablets is heavily fragmented, yet the key events can still be pieced together from what is left.

Of course, the Hittites did not know what it is they were preserving for us, and they dressed up the leftovers of memory, that was about 7,000 years old at that stage, as a thriller full of intrigues, treachery and heroism suitable for the time. Once we get rid of the dross, we are left with several interesting facts:

1. The monster stone boy named Ullikummi grew up on the shoulders of a Hittite version of Atlas, whose job was also to hold up the sky. In our parlance we would say that, in the distance, an asteroid was detected in space and, as it moved closer, it 'grew' in size.

2. The stone boy's task was to destroy the gods' cities in the sky, and the story obligingly provides three names. I.e. the sky people realized the asteroid is going to impact the three surviving satellites.

3. The gods are petrified at their fate, but decided to fight this monster "before the skull of head becomes overwhelming." The fight though was not easy, as the leading god tells his underling: "first I struck him, the Stone; now go ye and fight him again." I.e. destroying such a huge asteroid obviously was not easy, even for the sky people, and the best they could do was to keep breaking it up into smaller and smaller pieces.

The outcome of this 'story' is probably recorded in Hesiod's *Theogony*, in the part where Zeus fights Typhon, on whose "shoulders grew a hundred snaky heads," and whose "eyes flashed fire." Filling in the blanks, we can assume the above is what became of the description of the asteroid, shattered to hundreds of pieces, streaming through the atmosphere, preceded by two large red hot fragments.

The satellites avoided being smashed by the large asteroid, but were most likely unable to avoid a near destruction from the impacts of the many large fragments, and countless smaller ones, which would have overwhelmed the satellites defense systems. At least two of the three space habitats were seriously enough damaged to be evacuated. Furthermore, it was most likely at this stage the two space habitats really started to fall towards the Earth, and eventually crashed there, because the 'myths' provide more circumstantial evidence for this scenario, as we shall presently see.

The Indian lore tells us that during the supremacy of the Dragon, that is during the reign of the Alfa Draconis constellation, indicating the North Pole between 3492 BCE and 1793 BCE, the majestic mansions of the lofty city of gods cracked and crumbled. Since the final destruction of the space elevator can also be dated to the same time period, and since there are hints of impact against the space elevator, we can assume that at least one of the falling space habitats crashed into it, after its orbit deteriorated. As another Chinese story implies, the space elevator was again severed from its mooring during the cataclysm that ended the Ice Age, and it started drifting in space above the planet. It is even possible it was the less damaged of the two space habitats that crashed into the space elevator, according to my interpretation of the relevant 'myths'.

Regardless of the details, there is an evidence of mass migration from space to Earth. It took place about 10,000 years ago, either in escape pods, or in gliders of some sort. The lore of the Peruvian Indians states their people were 'born' from bronze, gold, and silver eggs that have fallen from heaven. Among the North American natives exists this story: "The first Dakota was formed from a star; he hurled him and watched him as he fell through the darkness, until he rested on soft soil. He was not wounded, Wa-Kin-Yan, the first Sioux." In Tibet they say that in ancient times people swooped down out of the sky, while the pictures shown in the age old book in Lhasa are said to resemble UFOs. It is not impossible that such a mode of transport was also used, but as the story of the Ham tribes, from the Tibetan-Chinese border, suggests they were most likely a type of emergency escape pods: "The Dropas (sky people) came down from the clouds in their gliders." It is also quite possible that these escape pods were equipped with landing retro jets and landing struts, as a part of a Chinese story attempts to explain: "Brilliant sparks and golden feathers burst into the air like some strange celestial fireworks, and when that Sun hit the Earth, it turned into a strange three-legged crow." The Haida of British Columbia also maintain that "… great sages descended from the stars on discs of fire."

But many of these people were not as lucky as the first Sioux, and others, to land softly on the Earth. The Amerindians have quite a few stories about 'baskets' full of people crashing to Earth, but the Wintuns of California have perhaps the most telling one. This "basket as big as a house and (with) a rope which reached up to the sky … burst open; all came out, poured down, a great stream of people, and all fell straight into the fire of the sweathouse." All that was left of them was "but a heap of bones and ashes." Yet, there were survivors as well. "We have found a man, but he is all bones," a girl informed her father. "Take good care of the stranger ... he is a good man," replied the father. The man begun to look better, but he cried all the time," the story tells us. And so did other such people elsewhere, about the same time, aware of the unspeakable loss they have just suffered, and the implication of it for the future; their own, and of the entire human race.

In the Welsh poem, *The Spoils of the Nameless Place*, we can read the following lines: "Thrice the fullness of Prydwen (the magical ship) we went into it; except seven, none returned from Caer Sidi." The same tragic incantation is being repeated for each of the Revolving Castles. The level of the tragedy is explained by this line: "Three times twenty hundred stood on the wall." That means 6,000 people were left to die aboard each space habitat. "Except seven, none returned ..." My understanding of this poem is that the last spaceship, or 'glider', akin to the space shuttle, was able to save only a handful of people. Counting seven people from each of the eight revolving castles, as this poem does, the total would be fifty-six. Yet, I doubt such a conclusion to be true, because the references of similar escapes are simply too numerous, unless the very last 'glider' was meant. A complimentary Chinese story, allegedly departing in the opposite direction, for heaven, says only seventy-one people were able to save themselves on the back of a divine dragon. The details though, including the fact the event is taking place among sacred mountains, invalidates the claim of departing for heaven, but suggests the Earth was the destination.

Then there is the Hopi tale of Sipapuni – the Hole in the Sky, which we call the space elevator. The story tells us how the Spider Grandmother led the people 'up' through the three worlds. Or rather, following the roles of reversal, people from the three 'worlds', i.e. the space habitats, were conducted down to Earth. "Then they were on their own," the story tells us. A stark truth indeed, because, as the narrative makes it quite plain, all these people ended up inside the space elevator, represented here by a bamboo stalk. But "the entire stalk (is torn) from the ground below ... and as the magical bamboo crumpled, thousands of doomed human beings toppled like seeds ... into the darkness of the world below."

Such statements make one think of the many refugees from The War, who made their home inside the immense structure of the space elevator, where they then lived for millennia. But here is the Chinese 'myth', we have hinted at earlier, which addresses the rebellion of water, and how it "retreated to a height called Imperfect Mountain, where the other elements laid a siege of him. Finding the position hopeless, water despairingly struck its icy brow against the mountainside. Imperfect Mountain split in half, a deluge poured forth, the Pillars of Heaven tottered ..."

The above fairly believably describes the conditions in the Arctic Ocean during the peak of the last Ice Age, as well as during the cataclysmic termination of it that ripped the space elevator apart not only near the Earth's surface, but the entire "pillar cracked and fell in pieces, broken ..." as another Chinese story tells us. However, even if the Earth anchorage was crushed by the ice, the space elevator would not really crack and fall to Earth, but drift along in space as noted earlier – until it collided with the aforementioned descending space habitat(s) – and then disintegrate. The fragments, or their remnants, would eventually crash onto the surface. During either or both of these events a large number of people would have perished. A Borneo 'myth' states that mankind sprang from a living rock in the middle of the sea, suggesting that representatives of many races have lived for many generations inside the Earth-bound portion of the space elevator, and emerged from it unscathed. We can only speculate whether this refers to the situation before the termination of the Ice Age proper, or after the termination.

The Piutes believe "the Amerindians were created in the sky by Gitchie Manitu, who sent down here a big thunderbird to find a place for his children to live." This story not only confirms the account of hybridization of the sky people and of the surface people, but in a sense confirms it twice over. You see, the 'myths' also state the sky people were of the 'white' race. That, of course, does not explain much at this point, but bear with me for a while longer. First we need to establish the fact.

For starters, the natives of the Gulf of Guinea, Africa, say: "White men had once come from the sky and from an island (there), now no longer in existence." In the Mayan sacred book *Chilam Balam* it is written: "Creatures arriving from the sky on flying ships ... white gods who fly above the spheres and reach the stars." (Note:

They do not come from the stars, they go there.) And in a Vedic text we can read: "The path of Aryaman" (i.e. the white man) "is a path leading from a star to the Earth."

The clincher?

Mitochondrinal evidence found in about 10 percent of the Amerindian natives confirms two infusions of 'Caucasoid' genes. The first such infusion took place, geneticists estimate, between 25,000 to 20,000 years ago, which would point to the period of the genetic 'adjustment'. This, in turn, confirms the date for The War, which would have taken place somewhat earlier. The second such infusion took place about 10,000 years ago, which again confirms the time frame offered above. But this time the cross-breeding was taking place naturally, on the surface of the planet, following the escape of the sky people from their disintegrating space habitats. As already indicated, some of the escape pods, or fragments of the space habitats, with survivors in it, are reported not only from the Americas, but also from the Eurasian continent.

By the rule of averages, most of the fragments should have ended up in the open seas, since they cover most of the Earth's surface. But it is also possible the sky people had managed to 'aim' the falling space habitats to impact the open seas near a coast, which, in relative terms, would 'soften' the impact, and increase the chances of survival for some. In either case, indications are that at least one space habitat crashed in a sea, scattering burning debris as it fell.

In the Finnish epic *Kalevala*, we can read an account of the moment when one of the space habitats, till then circling the planet in the upper reaches of the atmosphere, possibly after it grazed, or impacted the Space Elevator, was grabbed by the Earth's magnetic field, and was being pulled down: "On the air-merge sat the maiden, there it was she rocked the fire ... and the stupid maiden dropped it ... from her hands the fire dropped downward ... then the sky was cleft asunder, all air was filled with windows burst asunder by the fire ... and the red drop quick descended ..."

It is also remotely possible that the people inside were somehow able to slow the structure down, just before the impact, or at least of parts of it, jettisoned for such a purpose, as the following account seems to suggest. But it could also be merely wishful thinking that permeates much of the story of the Wheel King we mentioned earlier. The story concludes by saying: "... the Celestial

Wheel, plunging down into the Eastern Ocean, rose up out, and rolled onwards to the region of the South ..." Such an image though also invokes the idea of the proverbial "dead-cat-bounce," or of painfully naïve imagination that made the cyclopean Wheel roll on water, as an oversized skipping stone.

Probably the most realistic rendition of such an impact is again contained in the *Kalevala*, in the section devoted to the Mother of the Waters, also known as the Halcyon Bird, who laid six golden 'eggs,' and one of iron – a yet different image for the six space habitats and the space elevator. After three days, (and here a day equals about 100,000 years), the 'eggs' started to get hot and hotter. She "... rolled the eggs into the water, down amidst the waves of oceans, and to splinters they were broken, and to fragments they were shattered."

Yet, we do know there were survivors, but as the lamentations recorded in the Mayan sacred book, the *Popol Vuh*, indicate, that to endure living on the planet, under the primitive conditions they were plunged into, was quite unbearable for the sky people, tended to all their lives by robots: "And there they broke apart. There were those who went eastward (from Tulan) and many who came here, but they were all alike in dressing with hides. There were no clothes of the better kinds ... but they were people of genius in their very being when they came away from Tulan Zuyua, Seven Caves, Seven Canyons so says the Ancient Word ... In unity they waited there for the rising of the great star named daybringer ..." (I.e. Venus) "They cried their hearts and their guts out ... and they looked terrible. Great sorrow, great anguish came over them; they were marked by their pain. They just stayed that way. 'Coming here hasn't been sweet for us. Alas! ... What have we done? We all had one identity, one mountain, but we sent ourselves into exile ... this very place has become our mountain, our plain.'" (That is from the inverted "mountain-plain" of the space habitat to the surface of the planet.)

The account continues by describing how the survivors were reduced to eating larvae of insects, and eventually to cannibalism, and concludes: "It was actually a long time before the tribes came to their senses." And they suffered this insurmountable shock in spite the relevant knowledge provided for them by "the ('stone') gods (who) taught procedures," or the variations on the "teraphim," that is the wearable computers we have already encountered, which the

Mayans called Tohil, the Seneca Indians called them the Storytelling Stones, and the Hebrews called them Cherubim, which means "full of knowledge."

The lamentation though was not just about their personal tragedy and debasement. The Mayan text specifically mentions that those who were too old, or too young, had to be left behind, but also about their crushed egos, inflated by the eons of virtual godhood, which blinded them to the basic fact – they rebelled against the plan and will of the Supreme Being, and in their blindness, made worse by their 'power', they thought they could go on resisting indefinitely the Supreme Being's evolutionary goal for the life on this planet. They only slowed down the process, while setting themselves up for this dreadful fall. Furthermore, they obviously had no survival skills, or manual dexterity.

While it is true that the sky people used to intervene on the planet when required, most likely through their Earth-born agents, to rectify the worst ecological calamities brought about by human activities or by natural events, and they almost certainly averted impacts of several asteroids upon this planet, but they steadfastly refused to return to the planet to accept their fate – to be "surface people." Of course the conditions here were often 'unsettled', but with their knowledge and technology, they could have changed that as well. It is also true that people on the planet, the way they were, most likely represented a major problem. But, as mentioned already, the cataclysm that afflicted the planet about 100,000 years ago, left only 10,000 people alive, if that many. The population of the space habitats must have been much, much larger than that, and they could have repopulated the Earth practically all by themselves, and establish a culture here that would suite them. That, of course, implied a certain amount of difficulties and hard work, at least now and then, which the sky people did not feel like putting up with, preferring their life of ease and comfort while virtually all their work was done by the robots. That was not the case on the planet, because to maintain such a lifestyle here requires a lot of energy and resources, and has to be limited to a fairly small, steady population. Furthermore, if the population was to grow on the planet, as it would have to, in order to be effective in its task, the problems and a perceived inequality would grow as well.

And there was still another crucial reason that kept the sky people where they were: with the fluctuating temperature on the surface of the planet the size of the surface people's craniums also fluctuate, along with the corresponding mental capacity. A factor they were able to eradicate inside their habitats. What they did not realize, or did not want to, was the implication that the implicit goal of the Guided Evolution was the optimum size and shape of the human cranium, capable of an optimum performance even under the most adverse climatic conditions. They kept refusing to acknowledge the fact, and do something about it – until forced to it by very dire circumstances. Ultimately, most of them ended up paying dearly for their arrogance, while at (almost) the same time their personal genious significantly contributed to the completion of the Guided Evolution.

The theme of the rebellion of Lucifer and his angels, and of their fall, is known to us all. Yet, this fall is viewed with a certain sorrow even in the Bible: "How are thou fallen from heaven, O Lucifer, son of morning. How are thou cut to the ground, which didst weaken the nations! For thou hast said in thine heart: I will ascent into heaven, I will exalt my throne above the stars of God; I will sit also upon the mount of the congregation, in the sides of the north; I will ascend above the clouds …"

This excerpt, from Isiah 14:12-14, tells us a few things not readily noticed by the Bible readers: Lucifer was deemed to be the 'son' of the morning star, which is Venus, as were the Mayan ancestors. But not of the Venus as we think of it today, but one of the space habitats mistaken for it. Originally, Lucifer was seen as the bringer of light, and in this respect he can be identified with Prometheus, who was also a bringer of light, in form of solar, or mental fire, as well as a teacher of mankind. He too ended up being punished. His lot was to be "chained to a mountain." As we saw, the theme of the 'mountain' popped up even in the short quote above, Also, as we noted previously, the word "throne," that was believed to refer to a specific group of angels, should be read as "rotating." In general it can be said that Prometheus (Foresight), a.k.a. Lucifer, represented the oldest generation of sky gods, i.e. of sky people. (The very name Lucifer, composed of "Lucy" and "fairy" means an aerial being of light.)

The 'Book' of Obadiah, also found in the Bible, provides us with additional relevant material, once we remove the accrued gobbledygook that is attempting to make it look 'historical': "The pride of thine heart: who shall bring me down to the ground?

Though thou exalt thyself as the eagle, and though thou set thy nest among the stars, thence will I bring thee down, saith the Lord ... even destroy the wise men out of Edom, and understanding out of the Mount of Esau ..." While the Bible tells us Esau had a twin brother, it does not tell us that Esau was a "mountain," a "twin" mountain in fact, the counterpart of the "twin nest" of the Egyptians. By the same token, the story of Gilgamesh does not tell us either that "Gilgal" used to mean a "wheel" or a "revolving sphere," and "Mashu" a "twin mountain." In other words, Gilgamesh used to mean a "twin revolving mountain." A bit of a misnomer, since the "twin mountain" is the space elevator, connecting the twin worlds, as already was explained earlier, when we discussed the Japanese 'twins'. The implication then is Esau originally represented the space elevator. However, the spiraling red orbiter, which also had it 'up' and 'down' portion, could also be deemed a 'twin, revolving mountain' as well.

The Gnostic texts though are somewhat more specific about the real causes of the downfall. They explain to us that "the prime parent became insolent ... He said 'It is I who am God ...' " after "all the gods and their angels gave blessing and honour to him." Of course, this 'parent' was the same one who created the six heavens, or a generic name for all the sky people. "Then when Pistis (Faith) saw the impiety of the chief ruler, she was filled with anger. She was invisible." (This in fact says Pistis represented the Supreme Being.) She said, "You are mistaken, Samael," which means a blind god. As a punishment, she told him, "you will descent to your mother, the abyss, along with those that belong to you. For at the consummation of your" (plural!) "works the entire defect that has become visible ... will be abolished, and it will cease to be and will be like what has never been ... and ... heavens will fall one upon the next and their forces consumed by fire ... and his (Yaltabaoth's) heaven will fall and break in two."

Aside from the hint at the probable causes, the rest of the images are rather familiar: a mass destruction of the space habitats, falling in flames into the 'abyss', i.e. to Earth, the mother. Again Yaltabaoth's 'heaven', i.e. the space elevator, is broken in two, and falls as well.

A similar sentiment can also be found in the Egyptian *Book of the Dead*, whose origin goes back about 5,000 years. There it is Horus, the Sun-god, i.e. the god of light, (re: Lucifer, Prometheus,

Ahura Mazda, etc.), but also acting the role of the Supreme Being, who threatens: "I shall annihilate the devils; those who travel through the sky, those who live on Earth, and also those who reach the stars." The following words were also put into Horus' mouth: "I approached the accursed zone into which they had fallen ..." (I.e. the Earth.) "In truth they could not retrace their orbits of old because their path is destroyed." I.e. they lost the ability, the "know-how," to ascend.

There is one last argument to be made against the sky people: they most likely represented a collection of the brightest and most creative people, with the greatest organizational abilities, who have ever lived on Earth, before they relocated to their space habitats. Their departure though has left the human species intellectually impoverished. As the Gnostics said, the masters kept the world "restrained by ignorance." For that too they had to pay, by (re)experiencing it themselves.

The entire history of the last 300,000 odd years is presented, symbolically, by a three story pyramid, whose remains were found, in 1725, by a French explorer, Father Duprac, in northern Tibet, constructed by the Hsing-Nu, a non-Chinese people, who used to live there in the ages past. The Tibetan monks told him the three tiers represent three stages of mankind. The bottom one represents the Ancient Land, where men rose to the stars. The middle one represents the time when men came down from the stars. The top represents the New Land, the world of distant stars, *now* distant and inaccessible. It is also said of these Hsing Nu that they became engulfed by a "fiery cataclysm" and the survivors lapsed into barbarism and superstition. Exactly the state of affairs they probably feared the most.

We shall now attempt to describe what effect the demise of the space habitats had on the affairs of the people on the planet. But in order to keep things as clear as possible, I have to, at this stage, get ahead of myself and the timeline, all the way to the 7th century BCE. It was between 787 BCE and 687 BCE that the calendars all over the globe were suddenly in a chaotic state and calendar changes were being implemented. In the 8th century BCE in Egypt, and in the 7th century BCE in India and China, as astronomers realized that their ideas about the position of planets in the sky did not correspond with the images they could see there.

The explanation of what had happened is partially to be found in myths, and we shall attempt to recover it here as much as possible. Not because of an inordinate interest in astronomy, or because I would think the ancient people were exceptionally interested in it, but because it provides a yet another proof for the former existence of the space habitats.

The Mayans, whose ancestors most likely came from the space habitat originally identified as 'Venus', date the 'birth' of that 'planet' precisely to August 10, 3113 BCE, because they most likely left that habitat on that particular day.

Chaldeans of the 2nd millennium BCE, the inheritors of the Sumerian knowledge, were aware of Venus, as well as of Mercury, Mars, Jupiter, Saturn and Uranus. But the succeeding generations of Mesopotamian astronomers claimed knowledge only of Mercury, Mars, Jupiter and Saturn, as did the Brahmans in India, according to the ancient Hindu *Tablet of Planets*, dated to 3,102 BCE. "Venus alone is not found," informed the readers J.B.J. Delambre, in his *Historie de l'astronomie ancienne*, published in 1817, referring to that fact. But, if they could see Mercury, next to the glowing Sun, how could they have missed Venus, usually the brightest of all observable planets?

Of course, they could not have failed seeing it, but they did not recognize it for what it was, because to them the planet was 'invisible'. But this is not a silly joke, because the Chinese astronomers were well aware of the "invisible planets circling the Earth." These "invisible planets," of course, were the abandoned space habitats, and broken up sections of the Space Elevator. This would also explain why the Hindu astronomers mistook the real Venus for an anonymous star.

But then, all of a sudden, the world was bursting with dramatic news of this 'planet'. Even the Tahitians, Innuit, Yakuts and Kirghis started to talk about the 'birth of Venus', while implying something else than did the Mayans. The Samoans observed: "The planet Venus became wild and horns grew out of her head." The Venus Tablets of Ammizaduga, found in the ruins of the Assurbanipal Library in Niniveh, and dated (for now) to the first Babylonian Dynasty (1830 – 1530 BCE) have this to say: "On the 11th day of Sivan, Venus disappeared in the west, remained absent in the sky for nine months and four days, and on the 15th of Adar she was seen in the East." The next year this 'Venus' was gone for two months and six days; a year later, for eleven days and a year after that for five months and sixteen days, etc. Needless to add, no real planet in our solar system could behave like that.

"... the brilliant star Venus ... changed its color, size, form, course, which never happened before nor since. Adratus of Cyzinus, and Dion of Naples, famous mathematicians" (of antiquity) "said that it occurred in the reign of Ogyges," wrote Marcus Varro in his book *Of the Race of the Roman People*. While the reign of the mythical Ogyges coincides with the occurrence of the latest (major) global flood, and thus can be dated to the 24th century BCE, the calculations of the two mathematicians could be questioned, as could be the Babylonian date. Still, the two 'extreme' but vague dates more or less define the period to somewhere around 2,000 BCE.

Interestingly enough, just around 2,000 BCE a large asteroid is said to have impacted in central Sahara, but before doing so, its burning debris incinerated forests in India, Arabia Felix, Egypt, and the Sahara itself, wrecking in the process the Mohenjo-Daro civilization, and obliterated, among others, the cities of southern Mesopotamia, along with Sodom and Gomorrah, in Trans-Jordan. Ovid noted that "Libya," a general term for northern Africa at the time, became a desert in the wake of this asteroid's impact, known to the Greeks as Phaeton, who drove his father's, that is the Sun's chariot, so close to the Earth, the fable says, he set it afire. His 'joyride' was cut short, we are told, by Zeus' (read also Jupiter's) bolt of lightning.

Now to backtrack somewhat. As far as I know, no sizeable crater from that time period (2,000 BCE) was reported found in the Sahara. I would suggest this 'asteroid' did not create any such crater, because, being hollow, it lacked the appropriate 'punch.' It was hollow, because it was no other than the space habitat known to the earthlings as 'Venus'. As we have often noted, such space habitats were, among other names, known also as 'mountains'. When such a 'mountain' crashes on a planet, it is buried under itself, or within itself, so to say, in reference to the Typhon, most likely another version of the same event mythologized by the Greeks, and, at the same time confused with a similar event that took place circa six millennia earlier. Furthermore, the very components of the name Typhon suggests that it was melting, dissolving, while falling.

The murky picture starts to clear up with the help of historians who tell us that Typhon, in the Egyptian mythology, was no other than Seth who fought his brother Osiris, and in the end cut him up to fourteen pieces. This very struggle has, in turn, its counterpart in the Greek lore, albeit under different names. In one of the Homeric hymns we can read: "Most mighty Ares ... chieftain of valour, revolving thy fiery circle in ether among the seven wandering stars, where thy flaming steeds ever uplift thee above the third chariot."

Ares was the Greek war god, before the Romans renamed him Mars, and the same way this 'god' was identifiable with the red planet Mars, we can identify him with the 'red' or 'bronze' space habitat. Maybe yet, while the anonymous Greek poet has attempted to identify for us the right space habitat, he was obviously confused about the state of affairs in heaven, because he talks in the same breath about "seven wandering stars," i.e. space habitats, while at least three were long gone by then, but at the same time he says "ever uplift thee above the third chariot." Here I would suspect the "third chariot" was most likely the main section of the space elevator, still floating in space, or one of the other remaining space habitats, also slowly falling to Earth.

The explanation for this part of the puzzle is offered, at least to a degree, in the following quote from Homer's *Iliad*: "Ares ... first leapt upon Athena ... (and) smote her upon her tasseled Aegis ... but she gave ground, and seized with her stout hand a stone ... black and jagged and great ... therewith she smote furious Ares on the neck, and loosened his limbs." Aphrodite came to aid the wounded Ares, but Athena "smote Aphrodite on the breast ... and her heart *melted*." (My italics)

What does it all mean?

Within the given context, this is the most plausible scenario, I can offer: Two of the three remaining space habitats, knocked off from their L1 point, during The War of the Gods, by the explosion of the other three space habitats, descended low enough, by around 4,000 years ago, to come within a 'striking distance' of the wrecked space elevator. It most likely was after one of them, the abandoned one, designated as 'Venus', was further damaged either by a space debris, or by a grazing collision with the red space habitat, because the text notes specifically "Ares ... smote her upon her tasseled aegis ..." While aegis was Athena's famous shield, the stress here is on the word "tasseled." It implies something "tread-like" or "veil-like" an expression not dissimilar to the image of smoke offered by the Mayans, who called Venus "a star that smoked." (The Chinese said: "Venus was visible in full daylight, and ... rivaled the Sun in brightness.") The Chaldeans called it "a bearded star," and a "bright torch or heaven." Of course, this Venus was Athena as well, but also Ishtar, of the Assyrians. In another Homeric hymn, to Athena, we find one other name for her: "the glorious goddess, virgin Tritogenia." Yet, she "is reported to have appeared ... in the times of Ogyges," claims St. Augustine, echoing Varro in *The City of God*, reconfirming for us the date of its appearance, and of its subsequent disappearance. (Tritogenia most likely means the "third queen," possibly referring to the three space habitats.)

But to finish our 'scenario': after one of the space habitats, identified as Ares, passed through the screen of smoke, or flame, that was trailing behind 'Venus', it in fact grazed or impacted the massive remains of the Space Elevator, the "Great Stone." According to another version of the story, Athena drove a spear into Ares' body. This could be suggesting that a part of the space elevator passed through the ring of the satellite, damaging the support structures, without destroying it outright. This, in turn, is confirmed by the *Soochow Astronomical Chart*. It states that Venus once ran off the zodiac and attacked the "Wolf Star." But what the Chinese called the "Wolf Star," the Egyptians called the "Dog Star," was in fact the Space Elevator.

The relationship between 'Sirius' and the six space habitats is established, interestingly enough, by the story of the giant Orion, the Moon-Man-of the Mountain, chasing after the six Pleiades. To save them, Zeus, we are told, turned them into stars. Sometimes later the said Orion was killed by Artemis. The excuse given is the giant was walking too far out in the 'sea' to be clearly recognizable, because only his 'head' was seen above the 'waters'. Stripped of its mythologized verbiage, the 'sea' becomes the "River Ocean" above the Earth, and the Orion's giant head becomes a giant topmost fragment of the space elevator, by then probably shattered into many pieces. Yet, as far as their function is concerned, we can place an equalization mark between 'Orion' and 'Sirius' as well, but also between the two and the 'Venus'. And the ancient Egyptians claimed that the 'union' between Sothis, a.k.a. Sirius, the Dog Star, and Osiris, a.k.a. Orion, i.e. two parts of the space elevator, gave birth to 'Venus." In Babylonia, Saturn, Jupiter and Venus were, in that order, all given the name of Baal. This remnant, I strongly suspect, was also known as the "Golden Fleece." At least along the Guinea Coast of Africa they have a story of a stormy ram whose fleece was fiery gold. He also had white-hot horns and flame for breath. He was so destructive that the people "who were powerful in those days, thrust the ram ... far away into the sky."

If the above seems 'somewhat' confusing, it is because the 'facts' are confusing, and while I am doing my best to explain it as clearly as I can, it still remains difficult to explain so many unclear, conflicting observations that have been given so many names. If you are still unsure though, reread the relevant paragraphs, or just ignore the details.

The space elevator was also a part of the original story that became the Greek myth of the *Golden Fleece*. The original meaning of the name given to the ship Argo was "White Metal," and most likely represented a story of a maintenance vessel, dispatched to repair the severed structure. Seen like this, the "clashing rocks" of the *Iliad* would represent the two parts of the space elevator, clashing against each other.

But it could also be a description of a malfunctioning iris-type gate inside the severed upper part of the Space Elevator. This possibility seems virtually confirmed by several versions of a myth, indigenous to the west coast of Canada. In the Haida and Tsimsyan renditions, it is not a ship, but a Raven, or a great chief's son in his father's Raven cloak, who flew to the Chief of the sky for "the ball of light (to) restore the light to the world." Instead of sailing up the Dardanelles though, the Raven "traveled up the Nass River ... until he encountered a very high mountain, which was too high to fly over. He ... saw that the passageway through this mountain opened and closed alternately. He watched it for a long while ... saw that it was very bright at the far end ... (and) as soon as the passage opened, the Raven flew through it, and just as he got through, the passage closed ... nearly crushing him."

Furthermore, since so many stars were given the name Argo in antiquity, we can also assume the traffic around the free-floating part of the space elevator was relatively busy, for repair and other purposes.

As hinted earlier, we could also suspect that the space elevator, seen above the planet, was even mistaken for 'Zeus' zapping 'Phaeton,' or 'Ares' for that matter, with a 'thunderbolt,' as massive electrical discharges took place, when these huge bodies passed close to each other in space. Needless to stress, it would have been all but impossible for the contemporary observers to guess which structure was which, when these came together amid smoke and fire.

In summing up the fate of the space habitat that became known as 'Venus', we can say that it was first most likely severely damaged, along with the other space habitat(s), by the fragmented asteroid, around 10,000 years ago, and both had to be eventually abandoned. Of the two damaged space habitats, the one designated as 'Venus' most likely also suffered a complete power failure, and disappeared from view. This event is probably captured in the story of the Tlingit Indians, from the northwest corner of Canada: "Now I am going to shoot that star next to the Moon," thought the young archer. "In that spot was a large and a very bright one. He shot an arrow at this star ... when, sure enough, the star darkened."

We have identified this 'star' as Caer Arienrhod, which most likely was 'snail,' or spiral shaped, and featured the largest surface of the three remaining space habitats. It almost certainly is the "tiered mountain," encountered by the little war gods we remember from the Hopi story, and the same one where, as the story from New Hebrides tells us, were found all the abandoned robots. It was dark inside, remember?

Because its surface was so large, and its defense systems were no longer functional, it is quite likely this abandoned space habitat was impacted by another large asteroid that ignited the chemicals still aboard, which started to 'smoke'. Divided by the rotating movement of this structure, the two jets appeared that were seen on Earth as a "pair of bull's horns," and were described as such by Samoans, Persians and Babylonians. Even the rabbinical authorities claim that the Hebrews, escaping Egypt, saw this phenomenon, while passing through the Red Sea, which somehow changes the date of their fabled departure.

The impact from the asteroid would have also sent the structure careening erratically around the nearby space, as described by the Babylonians. And, as recorded by Marcus Varro, the said space habitat, that could have been also mistaken for the space elevator, as it changed colour, size and form, depending on the angle it was seen from, which would have only added to the confusion. Near the end it must have been difficult to say who is Athene, or Venus, and who is, presumably "red" Ares, because at that stage none of these 'stars' were familiar any more. In her final phases 'Venus' became a "circling star which scatters its flame in fire," noted the Egyptians, as it entered the Earth's gravitational field. Then, as we can read in the *Iliad*, Palas Athene "darted down to Earth a gleaming star." The name Phaeton means "the blazing star." As the Egyptian priest told Solon, regarding Phaeton: "This legend has the air of a fable; but the truth behind it is a deviation of the bodies that revolve in heaven round the Earth." (Via Plato.)

The next to come down from the sky most likely were the remnants of the Space Elevator, possibly in the 8th or 7th century BCE. According to Sir John Frazier, Numa, an ancient royal legislator in Rome, tricked the rustic deities Picus and Faunus to draw down Jupiter from the sky! By charms and spells, how else. We shall ignore the simplicity of such an assumption, and be thankful for the pointer. It only confirms my conclusion that Jupiter, a.k.a. Zeus, indeed was one of the names for the space elevator, or what was left of it.

Yet, others ascribe the strange events in the sky of that time to 'Mars'. Two of the items that fell down were made into an ancile of the Roman Mars, and the famous Palladium of Palas Athene. In the Babylonian mythology of the time a dramatic shift has taken place as well. Nergal, originally a war-god, credited with "still(ing) the heavens," was accused of causing "the Earth to shudder." Worse yet, he "rushed" down and became the new king of the underworld. That is of the Earth, this time, which always was 'under' the Firmaments of the sky.

But there was more. During a war between the Egyptians and Assyrians, most of the Assyrian army succumbed to an unknown malady, sent by 'Hephaestus', the Egyptians believed. (I.e. Ptah.) Obviously something potently virulent fell from the sky just before the battle, and it fell on the Assyrian side. The date, fairly credibly calculated by Immanuel Velikovsky, was March 23rd 687 BCE.

On the very same date, as the following description attributed to Confucious attests to, there were extremely unusual events in the Chinese skies as well: "... during the night the fixed stars did not appear, though the night was cloudless. In the middle of the night stars fell like a rain." The plural "stars" strongly suggests they were in fact remnants of the space elevator, remnants thousands, even tens of thousands of kilometers long, that were floating above the planet for about 8,000 years, at that time. People have accepted these lights in the sky as 'stars', because they were bright, regularly circled the planet, and to the brightest ones the astronomers probably attributed names of the known planets and stars.

I have to thank again Mr. Velikovsky for digging out all kinds of 'odd' information to support his very flawed theory, but the facts, as the ancient people saw them, cannot be disputed. And according to the learned sources, Mr. Velikovsky is quoting in his *Worlds in Collision*, the Egyptians of about 3,000 years ago saw "stars shifting." Even the Orion constellation, so crucial for agriculture, and for ritual purposes in Egypt, started to shift, till 'Orion' appeared west of 'Sirius,' instead of east. These changes were also painted on the walls of the contemporary tombs. But then again, what they were looking at were not, of course, the real stars – in spite the fact the real constellation of Sirius was known to the builders of the Cheops pyramids, and to the Dogons of Mali in Africa. Yet, this 'sudden ignorance' is easily explainable by the upheavals people of this planet had experienced at that time, which made them forget writing in many parts of the world, as well as astronomy, for centuries.

Considering that the false Venus was 'replaced' by the real one, sometimes after 2,000 BCE, and that Saturn and Mercury were the other real planets in the skies of those times, the gradual 'disappearance' of Jupiter, Mars, and of the Sirius-Orion group, must have deeply disturbed the people, especially the astronomers. In the aftermath, as was already noted, the calendars were in chaos. But then, surprisingly fast, the real planets were located, and new calendars introduced, over a period of one hundred years, from 787 BCE to 687 BCE.

From South and Central America we have similar reports. The mythical Pachakum Inca is said to have "reorganized the stars," in the wake of a war. Also, the spherical stones, placed (usually) on mountain tops in Central America, are being associated with shifted celestial bodies.

The last space habitat probably came down in 235 BCE. Cicero, (106 – 43 BCE), referring to this event, speaks "of the time in which two Suns were seen ... and when the Sun of the night was seen, when noises were heard in the sky, and the sky seemed to split and strange spheres appeared in it." If the above still needs interpreting: the second "Sun" was the hot, rapidly descending space habitat, the "split sky" was probably the moment of explosion, and the "strange spheres" the escape pods.

Tito Livio (59 BCE – 17 CE) adds to the above: "The sky seemed to crack like a great wound, and across the opening shone a great light." Later, "in the calmness of the night, both consuls were visited, it is said, by the same apparition: a man of superhuman size and dignity who explained that he was the leader of one of the groups, while the other groups fleet had to be offered to Mani and to Mother Earth."

Reports of momentous events, such as the raining of milk and blood, on many occasions, of flesh, iron, iron 'sponges', wool, baked bricks and soil, that took place during the Roman times, were preserved for us by Pliny the Elder (23 – 79 CE), who also noted that a "comet was seen for seven days in the northern region" after Julius Caesar died, in 44 BCE. The ancient Chinese also reported blood and human parts falling from the sky at that time. But such incidents have continued till our times, and no one was better at recording them than Charles Fort. He reported on hundreds of such incidents, taking place in North America, Europe, and the rest of the world. The items that fell from the sky, often proverbially blue, include 500 tons of vegetable matter, "fish from heaven," some headless, some rotten, some still alive, and enough to cover 800 square meters, unknown types of grain, worms, large ants, stone axes (of the self-sharpening green stone), the customary blood, iron and steel, as well as iron slag and kiln fired bricks. Enormous chunks of ice also fell from the sky in the 19th century, long before the era of commercial airliners. Altogether thousands of frogs and birds fell from the sky, in sizeable groups, of many different species, some of them unknown. An iron "oblate spheroid," and a quartz disk five by six centimeters, five millimeters thick, fell from the sky as well.

Charles Fort also wrote: "Some notes I have upon remains of men and animals encysted, or covered with clay or stone, as if fired here as projectiles, I omit now ..." And it is too bad that he did, because he was on the right track. He even reached the inescapable conclusion: "The notion is that there is somewhere aloft a place of origin of life relatively to this Earth." At least he was a rational enough observer not to pretend he understood the nature of this lofty place, unlike the bulk of his inheritors, such as Velikovsky, Daniken, or Sitchyn.

As far as I am concerned, most of the phenomena described above are associated with the final destruction of the Firmaments.

The one remaining puzzle is the fate of the Revolving Castle the Celtic people called Caer Ochren, whose original name meant True Satellite. The name itself would suggest this space habitat still ought to be in orbit. The Gnostic texts, in their own convoluted way seem to agree, stating in fact that after The War (of the Gods) one of the Firmaments was "snatched up to the seventh heaven." The text implies that it was huge and four sided. And such a "four sided" or "four-cornered" space habitat was mentioned by the Celts and the Finns as well.

This "seventh heaven," if it is located anywhere specific, can most likely be identified with the Lagrangian point L3, situated behind the Moon, which would have blocked the view of it from Earth. Further measures, such as a "cloak of invisibility," would make it undetectable even to our instruments. After all, their technology is 300,000 years ahead of ours.

AFTERWORD

What was revealed in the preceding chapters amounts to nothing less than the fabled "secret of the ages." Over the millennia the 'secret' became so forgotten, practically no one knew what it was really all about, since at least 6,000 years ago, but likely longer. Never the less, the presented arguments, mutually supportive of each other, do demonstrate, beyond any reasonable doubt, that a highly evolved civilization, originating on Earth, had existed in the near-Earth space habitats, and that it was instrumental in its own demise.

Overall though, the sky people were, or rather still are, a highly successful species, and Evolution does not waste any successful species. It might transform them, and bounce them around somewhat, to teach them the required lessons, but they do not become extinct. At least not for a long, long time.

For example, the early reptilian species were transformed into dinosaurs, and these in turn into the current species of crocodiles, alligators, lizards, and, of course, birds of all sorts. There is evidence some brontosaurs, pterodactyls, and possibly plesiosauruses, have survived till historical times. In fact the odd brontosaur still might be alive in the swamps of Congo, Africa, according to the locals.

The same applies to the sky people. They are very much alive in our genes, and their presence there really baffles the current geneticists. They concluded, studying the mitochondria of our cells, which have their own DNA, that our species is only 250,000 years old, and that the entire species started out with *one* sole mother. Needless to say they are unable to explain this 'mystery', utterly impossible in the natural world. Especially since the same mitochondrial DNA is found in all the human races. The only rational way is to accept, as a fact, that the DNA of the surface people, alive circa 26,000 years ago, was 'blended' with the DNA of one specific sky-woman, who became known in the myths as the "birth-goddess." The status of the male DNA is less important as, by now, some 200 genetic 'Adams' were identified, and while it is also suspected of being capable of spontaneous mutations. The rest of the process, as even the Mesopotamian myth describes it, took place inside mechanical wombs.

But the sky people, now genetically compatible with the surface people, are still 'out there', although chastised, and much more willing to cooperate with the surface people, in the future, after our own evolution has been (almost) finalized. Till then, they have to keep their distance, in more sense than one, remaining at their off-world, or undersea basis. (And the same applies to any off-world visitors, which is the reason why we are not meeting any.)

But it was not always so. There is ample evidence 'suggesting' that the sky people were attempting to steer mankind in the right direction till the medieval times, revealing the relevant knowledge to selected individuals, or groups of individuals; especially where the conditions, in form of mental preparedness, were right for it, as they were in Western Europe. Unfortunately, many of the selected individuals were killed by the Church that accused them of practicing 'black magic', and of consorting with the 'devil'.

Realizing that, in their present reduced mental state, the new surface people are, by and large, beyond help, the sky people have resorted to a series of holographic messages from "the Lady," which the Christians presume is St. Mary, unaware that most of the goddesses representing the Earth used to be referred to as "the Lady." The proof of the holographic nature of these images was established in the 18th century Ireland, where people walked right through the one they were reverently observing.

Of course, no visiting aliens would interfere this way with the fate of the inhabitants of this planet, or have the cultural knowledge to do so, during a casual visit. Only the people, whose fates are intimately tied to this planet, and all the life here, would assume such a role; a role that was theirs, one way or the other, for the last 300,000 years.

Even if all six of the space habitats were destroyed, which is also a possibility, thousands and thousands of those sky people would have survived aboard the enormous space ships they had. The largest one we know of, mentioned by the Mayans, measured six kilometers in circumference. Since it "traveled to the stars," it must have been essentially self-sufficient in energy and food production, and able to sustain a population of thousands, if not tens of thousands, on such voyages. Also, the technology aboard must have been capable of not only renewing all the components of the ship, but of constructing a replica of it as well. Interestingly enough, in the mid to late 19th century, astronomers on several occasions saw sizeable flotillas of huge spaceships traveling through the solar system, some of them quite close to Earth. Asteroids are not known to travel together in "tremendous numbers," and cannot be said to be "self-illuminated," as most of those reported as spaceships were. Yet, while some of those objects reported, especially those near Jupiter, likely were asteroids, some of them, equally likely, were spaceships.

Enormous spaceships, complete with parks, zoos, hydroponic gardens, genetic banks, libraries, universities, etc. After all, the space-people would have learned their lesson from their role as sky people too: Do not live inside a large cumbersome, essentially unmovable structures, clustered in two vulnerable groups, orbiting the Earth. Scatter and survive. Live on the move like a hunter, rather than being the hunted prey, which also increases the chances of survival for Life on Earth.

Yes, we also have to accept, as a fact, that the Supreme Being 'hunts' the species that are not "nimble of thought and feet," because such do not fit the evolutionary needs, of maintaining life of this planet. As the new surface people, we will have to learn the same painful lesson, consciously, and apply it to living on the surface of the planet, cooperating with the new space-people, as well as with the other, older races, who, unknown to us, share the planet with us, living below the surface of it. And they have been living there about fifteen million years.

The existence of this subterranean race, or races, was also an accepted fact till the medieval times in Europe, and is fairly well documented in local stories, in the myths of North and South America, of Asia and the Pacific region. One can only speculate that these underground races, ranging in size from 'normal' to tiny, are somehow connected to regulating the activities of the lithosphere, the same way the space-race is (to be) responsible for protecting the Earth from undesirable impacts of asteroids/comets. Our task, as the future surface people, will be to maintain the optimum ecological conditions on the surface of the planet itself. Only this way can the human(oid) species, indigenous to this planet, become full participants in the adventure of life, embodying in a sense an insurance policy for the continuation of life here, regardless of any future perils the physical universe might pose. And it will pose a lot of such perils, which the Supreme Being will neither be able to avert, nor rectify fast enough, or at all; for example, when we leave the "local space bubble," in a fairly near future, as astronomers warn, and enter a region of space filled with dense dust and a lot of solid objects. Whatever work has to be done to protect all the life on this planet is strictly up to us now. In this respect we are not only the children of the sky gods, but also of the Supreme Being, and we have to carry out the task fated to us.

Otherwise there is nothing that could be summed up about the ultimate legacy of the sky people, except that it was completely forgotten. Of course, the ancient people have tried to keep it alive, either by their obsessive interest in astronomy, or by presenting the physical image of the space habitats, that is the 'wheel', during commemorative ceremonies, long before it was reduced from a holy symbol to its lowly task as a part of carts and wagons. The image of the Revolving Castles has also survived for a long time in form of medicine wheels, the wooden round castles of northern Europe, in ceremonial objects and decoration, and even in the edifice we know as the Stonehenge.

The various artificial mountains, such as the ziggurats and pyramids, which were the symbolical homes of the sky gods on Earth, also used to remind people of the artificial 'mountains' in the sky. The memory also took on the form of cultural images. For example, we know of the tradition of Chris Kringle, the Christ of the Burning Wheel, of the fall/winter fires atop the mountains of Old Europe, and, or course, the Creation story, or rather stories.

Several climatic swings, waves of mass migration, wars of survival, along with long periods of illiteracy, often centuries long, have left us nothing of that legacy, except a few scattered fragments, reconnected, for the first time in millennia, in this book.

INDEX

Aeschylus: PROMETHEUS BOUND – Dell Publishing 1965, New York, U.S.

Augros, R.M. & Stanciu, G.N: THE NEW STORY OF SCIENCE – Bantam Books, 1986, New York, U.S.

Barbeau, Marius: TSIMSYAN MYTHS – National Museum of Canada 1961, Ottawa

Benoist, Luc: SIGNS, SYMBOLS AND MYTHS – Victoria Publ. 1995 Prague, Czech Rep.

Bergier, Jaques: EXTRATERRESTIALS IN HISTORY – Odeon, 1992, Prague, Czech Rep.

Binder, Otto O: UNSOLVED MYSTERIES OF THE PAST – Tower Publ. New York, U.S.

Bohm, David: UNFOLDING MEANING – Ark Paperbacks 1987, London, U.K.

Bor, D.Z.: THE GOLDEN AND THE IRON AGE – Trigon, Prague 1985, Czech Rep.

Buck, William, tr.: MAHABHARATA – University of California Press 1973, U.S.

Budge, E.A. Wallis: EGYPTIAN MAGIC – Dover Publications, U.S. FROM FETISH TO GOD IN ANCIENT EGYPT – Dover Publ. U.S.,

Bowden, M.: APE MAN – Sovereign Publications 1977, Kent, U.K.

Campbell, Joseph: PRIMITIVE MYTHOLOGY – Penguin Books 1987, New York, U.S.

Chamberlain, Jonathan: CHINESE GODS – Pedaluk Publications 1987, Malaysia

Charroux, Robert: MASTERS OF THE WORLD – Berkley Publishing, New York, U.S.

Chetwyn, Tom: THE AGE OF MYTH – Harper Collins 1991, London, UK

Dalley, Stephanie, tr.: MYTHS FROM MESOPOTAMIA – Oxford University Press 1991, U.S.

Daniken, Erich, von: THE GOLD OF THE GODS – Bantam Books, 1974, New York, U.S.
: GODS FROM OUTER SPACE – Bantam Books 1972, New York, U.S.

Davies, Paul: GOD AND THE NEW PHYSICS – Penguin Books 1983, London, U.K.

Drake, W. Raymond: GOD AND SPACEMEN IN THE ANCIENT EAST – Sphere Books 1973, London, U.K.

Drummond, William: OEDIPUS JUDAICUS – Bracken Books, London, U.K.

Dundes, Alan: THE FLOOD MYTH – University of California Press 1988, U.S.

Dutt, Romesh: MAHABHARATA AND RAMAYANA – J.M. Dent 1917, London, U.K.

Eberhard, Wolfram: FOLKTALES OF CHINA – Pocket Books 1978, New York, U.S.

Eliade, Mircea: THE MYTH OF ETERNAL RETURN – Princeton/Bolingen 1974, U.S.: MYTH AND REALITY – Harper & Row 1988, New York, U.S.: THE FORGE AND THE CRUCIBLE – The University of Chicago Press 1991, U.S.

Eliot, Alexander: THE UNIVERSAL MYTHS – New American Library 1991, New York, U.S.

Feldman, Suzan, ed.: THE STORYTELLING STONE – Dell Publishing 1971, New York. U.S.

Fiske, John: MYTHS AND MYTHS MAKERS – Random House 1996, London, U.K.

Fort, Charles: THE BOOK OF THE DAMNED – Ace Publishing, New York, U.S.

Frazer, James: THE GOLDEN BOUGH – Gramercy Books 1981, New York, U.S.

Gantz, Jeffrey, tr.: THE MABINOGION – Penguin Books 1995, New York, U.S.

Goldsmith, Donald: NEMESIS – Berkley Books 1986, New York, U.S.

Graves, Robert: THE GREEK MYTHS – Penguin Books 1960, London, U.K.: THE WHITE GODDESS – Faber and Faber 1984, London, U.K.

Griavle, Marcel: CONVERSATIONS WITH OGOTOMELLI – Oxford University Press 1980, U.S.

Gribbin, John & Mary: ICE AGE – Penguin Books 2001, London, U.K.

Hamilton, Edith: MYTHOLOGY – A Mentor Book 1977, U.S.

Hancock, Graham: UNDERWORLD – Doubleday 2002, Toronto, Canada

Harris, Marvin: CANNIBALS AND KINGS – Vintage Books 1978, New York, U.S.

Heisenberg, Werner: PHYSICS AND PHILOSOPHY – Harper & Row 1987, New York, U.S.

Henderson, J.L. and Oakes, M.: THE WISDOM OF THE SERPENT – Collier Books 1971, New York, U.S.

Herodotus (David Green tr.): THE HISTORY – The University of Chicago Press, 1988, U.S.

Hruska, B. et al: MYTHS OF OLD MESOPOTAMIA – Odeon Publishing 1977, Prague, Czech Republic

Ivimy, John: THE SPHINX & THE MEGALITH – Harper Colophon 1975,
        New York, U.S.
Jindra, J.: FOOTPRINTS FROM THE FUTURE – Ameriguide 1995,
        Czech Rep.
Ke Yuang, tr.; DRAGONS AND DYNASTIES – Penguin Books 1993,
        New York, U.S.
Kitto, H.D.F.: THE GREEKS – Penguin Books 1986, London, U.K.
Knight, C. & Lomas, R: URIEL'S MACHINE – Arrow Books 2000,
        London, U.K.
Kolosimo, Peter: NOT OF THIS WORLD – Bantam Books 1973, New
        York. U.S. : TIMELESS EARTH – Sphere Books, 1974, London,
        U.K.
Kramer, S.N.: MYTHOLOGIES OF THE ANCIENT WORLD – Doubleday
        1961, New York, U.S.
Kratochvil, Z. & Bouzek, J.: FROM MYTH TO LOGOS – Herrmann &
        Sons 1994, Prague, Czech Rep.
Levenson, Thomas: ICE TIME – Harper & Row 1989, New York, U.S.
Lévi-Strauss, Claude: MYTH AND MEANING – University of Toronto
        Press 1978, Toronto, Canada
Lucretius, Carus: ON THE NATURE OF THE UNIVERSE – Penguin Books
        1961, London, U.K.
Magoon Jr. F.P., tr.: THE KALEVALA – Harvard University Press 1995,
        U.S.
Milne, Anthony: OUR DROWNING WORLD – Prism Press 1989, U.K.
Noorberg, Rene: SECRETS OF THE LOST RACES – Harper & Row
        1977, New York, U.S.
Over, Raymond Van: SUN SONGS – New American Library 1989, New
        York, U.S.
O'Flaherty, Wendy Doniger, tr.: THE RIG VEDA – Penguin Books 1984,
        London, U.K.
Pagel, Heinz, R: THE COSMIC CODE – Bantam Books 1984, New York,
        U.S.
Pauwels, L. & Bergier, J: ETERNAL MAN – Granada Publishing 1973,
        U.K.
Peat, F. David: THE PHILOSOPHER'S STONE – Bantam Books 1991,
        New York, U.S.
Perry, John Wier: LORD OF THE FOUR QUARTERS – Collier Books,
        New York, U.S.
Plutarch: ON THE DELPHIC "E" – Herrmann & Sons 1995, Prague,
        Czech Rep.
Potts, Rick: HUMANITY'S DESCENT – Avon Books 1997, New York,
        U.S.
Puhvel Jaan: COMPARATIVE MYTHOLOGY – The John Hopkins
        University Press

Rajagopalachari, C. tr.: MAHABHARATA – Bhavan Books 1986,
        Bombay, India
Robinson, James M. gen. ed.: THE NAG HAMADI LIBRARY – Harper
        Collins 1990, U.S.
Santillana, Giorgio de & Dechend, Hertha von: HAMLET'S MILL –
        David R. Godine 1977, Boston, U.S.
Sousek, Z.: BOOKS OF SECRETS AND WISDOM – Vysehrad 1998,
        Prague, Czech Rep.
Squire, Charles: CELTIC MYTH & LEGEND – Newcastle Publishing
        1975, U.S.
Steiger, Brad: WORLDS BEFORE OUR OWN – Berkeley Publishing
        1979, New York, U.S.
Sullivan, William: THE SECRET OF THE INCAS – Crown Publishers
        1997, New York, U.S.
Talbot, Michael: BEYOND THE QUANTUM – Bantam Books 1988, New
        York, U.S.
Tedlock, Denis, tr.: POPOL VUH – Simon & Schuster 1985, New York,
        U.S.
Temple, Robert: THE CRYSTAL SUN – Arrow Books 2000, London,
        U.K.
Thomas, Andrew: WE ARE NOT THE FIRST – Bantam Books 1973, New
        York, U.S.
Valee, Jacques: PASSPORT TO MAGONIA – Tandem Books 1975,
        London, U.K.
Velikovsky, Immanuel: WORLDS IN COLLISION – Dell Publishing,
        New York, U.S.
Waters, Frank: BOOK OF THE HOPI – Viking Press 1969, New York,
        U.S.
Weinreb, Friedrich: SYMBOLOGY OF BIBLICAL LANGUAGE –
        Herrmann & Sons 1995, Prague, Czech Rep.

Also articles in periodicals such as Scientific American, Discovery, Archeology, The Globe & Mail, and others, along with notes from sources by now forgotten, as well as the Indo-European Dictionary, compiled by Julius Pokorny, found in the Houghton Mifflin Canadian Dictionary, published in 1960.